T0325401

Water Harvesting for Groundwater Management

Challenges in Water Management Series

Editor:

Justin Taberham
Publications and Environment Consultant, London, UK

Titles in the series:

Smart Water Technologies and Techniques: Data Capture and Analysis for Sustainable Water Management
David A. Lloyd Owen
2018
ISBN: 978-1-119-07864-7

Handbook of Knowledge Management for Sustainable Water Systems
Meir Russ
2018
ISBN: 978-1-119-27163-5

Industrial Water Resource Management: Challenges and Opportunities for Corporate Water Stewardship
Pradip K. Sengupta
2017
ISBN: 978-1-119-27250-2

Water Resources: A New Water Architecture
Alexander Lane, Michael Norton and Sandra Ryan
2017
ISBN: 978-1-118-79390-9

Urban Water Security
Robert C. Brears
2016
ISBN: 978-1-119-13172-4

Water Harvesting for Groundwater Management

Issues, Perspectives, Scope, and Challenges

Partha Sarathi Datta
Delhi, India

This edition first published 2019

Registered Offices
John Wiley & Sons, Inc., 111 River Street, Hoboken, NJ 07030, USA
John Wiley & Sons Ltd, The Atrium, Southern Gate, Chichester, West Sussex, PO19 8SQ, UK

Editorial Office
The Atrium, Southern Gate, Chichester, West Sussex, PO19 8SQ, UK

For details of our global editorial offices, customer services, and more information about Wiley products visit us at www.wiley.com.

Wiley also publishes its books in a variety of electronic formats and by print-on-demand. Some content that appears in standard print versions of this book may not be available in other formats.

Library of Congress Cataloging-in-Publication data

Names: Datta, P. S., 1950– author.
Title: Water harvesting for groundwater management : issues, perspectives, scope, and challenges / Partha Sarathi Datta.
Description: Hoboken, NJ : John Wiley & Sons, 2019. | Includes bibliographical references and index. |
Identifiers: LCCN 2018046971 (print) | LCCN 2018052025 (ebook) | ISBN 9781119472018 (Adobe PDF) | ISBN 9781119472025 (ePub) | ISBN 9781119471905 (hardcover)
Subjects: LCSH: Water harvesting. | Groundwater–Management.
Classification: LCC TD418 (ebook) | LCC TD418 .D38 2019 (print) | DDC 628.1/14–dc23
LC record available at https://lccn.loc.gov/2018046971

Cover Design: Wiley
Cover Image: © iStock.com/NosUA

Set in 10/12pt Warnock by SPi Global, Pondicherry, India
Printed in Singapore by C.O.S. Printers Pte Ltd

10 9 8 7 6 5 4 3 2 1

Nature Should Be Respected
Not Commanded

This book is dedicated to my wife and son for moral support and encouragement, who inspired me to write down my experiences.

Contents

Editor's Note – October 2018

Great books often take some time to come together. Professor Datta and I first spoke in December 2014 and it has taken four years for his book to reach publication. Very sadly, Professor Datta passed away in August 2018, before he saw this book published. However, the book text was completed and his project has now come to fruition.

This is a book of which his family can be proud; it is a treasure trove of information that will prove incredibly helpful to water management students, academics and professionals worldwide. It is truly a fitting testament to his 40 years in the sector.

Justin Taberham
Series Editor
London, UK
Website: www.justintaberham.com

Series Editor Foreword – Challenges in Water Management

The World Bank in 2014 noted:
'Water is one of the most basic human needs. With impacts on agriculture, education, energy, health, gender equity, and livelihood, water management underlies the most basic development challenges. Water is under unprecedented pressures as growing populations and economies demand more of it. Practically every development challenge of the 21st century – food security, managing rapid urbanization, energy security, environmental protection, adapting to climate change – requires urgent attention to water resources management.

Yet already, groundwater is being depleted faster than it is being replenished and worsening water quality degrades the environment and adds to costs. The pressures on water resources are expected to worsen because of climate change. There is ample evidence that climate change will increase hydrologic variability, resulting in extreme weather events such as droughts floods, and major storms. It will continue to have a profound impact on economies, health, lives, and livelihoods. The poorest people will suffer most.'

It is clear there are numerous challenges in water management in the 21st Century. In the 20th Century, most elements of water management had their own distinct set of organisations, skill sets, preferred approaches and professionals. The overlying issue of industrial pollution of water resources was managed from a 'point source' perspective.

However, it has become accepted that water management has to be seen from a holistic viewpoint and managed in an integrated manner. Our current key challenges include:

- The impact of climate change on water management, its many facets and challenges – extreme weather, developing resilience, storm-water management, future development and risks to infrastructure
- Implementing river basin/watershed/catchment management in a way that is effective and deliverable
- Water management and food and energy security
- The policy, legislation and regulatory framework that is required to rise to these challenges
- Social aspects of water management – equitable use and allocation of water resources, the potential for 'water wars', stakeholder engagement, valuing water and the ecosystems that depend upon it

This series highlights cutting-edge material in the global water management sector from a practitioner as well as an academic viewpoint. The issues covered in this series are of critical interest to advanced level undergraduates and Masters Students as well as industry, investors and the media.

Justin Taberham, CEnv
Series Editor
www.justintaberham.com

Foreword

Globally, the recent past in human history has seen increasing demands on the Earth's limited water resources due to the ever-growing population, urbanization, agriculture intensification, and competition for economic aspirations. Use of groundwater resources has also changed dramatically due to emerging new technologies, mechanization, specialization, and government policies that favor maximizing developmental activities and production. Although these changes have had many positive effects, there have also been negative impacts, such as unplanned indiscriminate overexploitation of groundwater leading to decline in water level; degradation of land, soil, and water resources at an alarming rate; loss of biological diversity; indiscriminate waste disposal, etc. With global demand for groundwater projected to outstrip supply by 2030, many areas may face water insecurity. Groundwater managers have relied largely on supply-side infrastructural projects, which are environmentally, economically, and politically expensive, and also create political tensions due to intra- and interstate administrative and political boundaries.

Water harvesting has been practiced successfully for millennia in some parts of the world. The principle of rainwater harvesting is essentially to capture high-intensity rainfall causing potentially damaging runoff from areas of low water-holding capacity, and using this for plant irrigation and/or water supply in areas where rainfall is limited, uneven or unreliable with frequent dry spells. Management of groundwater resources includes ensuring its assessment and development, and protection from depletion and degradation, for maximizing the environmental, economic, and social benefits of its equitable availability, integrated with improving its supply and consumption efficiency, economic efficiency to sustain investments and operations, and waste management. However, water harvesting potential, in reality, remains largely unknown, unacknowledged, and underappreciated.

This book provides an overview of proven good practice in rainwater harvesting all over the world, which can be adjusted to the local context, and forms a practical reference guide for the integration of water harvesting technologies into the planning and design of projects. It will inform planners, decision makers, and practitioners in the field for use at watershed and river basin level. For the first time, this book attempts to uncover, collate, and make available current knowledge about water harvesting technologies, and the settings in which they tend to perform best. On a broader scale, the book is intended to facilitate and disseminate good practice in water harvesting to local and regional planners/advisors, rural development consultants, rainwater harvesting networks and communities, project managers, extension agents, and other relevant staff.

Partha Sarathi Datta
New Delhi, June 2018

Preface

In the twenty-first century, the projected global demand for water is expected to outstrip supply due to the impacts of rapid growth in population, urbanization, agricultural intensification, industrialization and economic aspirations. Increasing water availability in terms of quantity, maintenance of water quality, and water security has become a focus of increasing international concern. Generally, to meet increased demands for water, water managers have relied on large-scale, supply-side infrastructural projects, such as dams and reservoirs, which are environmentally, economically, and politically costly. However, with the vast majority of water resources being transboundary, crossing intra- and interstate administrative and political boundaries, and often being ignored or taken for granted, supply-side projects can create political tensions. Hence, during the last three decades or so, for managing the water demand–supply gap for survival of humans, animals, and plants, the importance of rainwater and groundwater in the development process has been increasingly recognized.

Flexible approaches to policies and planning all over the world in the 1960s–1980s resulted in a revolution in the agricultural sector, the largest user of water resources. In spite of these commendable efforts, much remains to be done to use rainwater and groundwater resources optimally and efficiently on a sustainable basis because availability of these resources is fixed and supply is diminishing very rapidly, generally due to increasing demand and abuse. Water demand management involves, in particular, improving water harvesting and conservation, use and reuse to ensure water availability during times of both normal and atypical conditions, through the communication of innovative ideas, norms, and methods for water conservation and encouraging society to reduce water consumption by changes in practices and attitudes towards water resources. To communicate the value of groundwater resources, it is essential to examine their occurrence, conservation, distribution, consumption, and use for domestic, agricultural, and other purposes.

In our rapidly urbanizing world, access to groundwater resources is not only inadequate in many areas but also highly unsustainable. Unplanned exploitation of groundwater has been alleged to be responsible for water scarcity in many parts of the world, although scientific knowledge to support such a contention is limited. Hence, effective sustainable groundwater resources management requires a clear understanding of the linkages between various resources in a natural setting. The large infrastructure market and growing need for water supply intervention is attracting the interest of international development agencies, multinational corporations, and investors. Private sector involvement as a proposed mode to address this problem appears to be politically contentious. As shortages of resources occur and technology develops, so do their technical, economic, and environmental complexities.

My attitudes have been formed and opinions shaped by the steady accumulation of evidence and examples during my 40-year professional career, undertaking extensive field investigations in the context of the Groundwater Recharge Assessment and Management, International Hydrological Program (IHP, UNESCO), the Asian Regional Co-ordination Committee on Hydrology, Agricultural Research and Development, and Environmental Impacts Analysis. I have had the opportunity to analyze national and international academic work and review reports and policy documents on natural resources in the political and historical landscapes underlying global water resources demand, supply, and consumption. I believe that all the alternatives and their potential costs and benefits need to be carefully evaluated based on sound knowledge of the resources and related aspects before any final decision can be taken. Continued scarcity and degradation of water complicated by the variability and intensity of weather patterns has become an increasingly serious concern.

For example, in India, due to increasing population and multisectoral demand for water resources, water supply is multidimensional, linked to reliable assessment of availability, scope for distribution, augmentation, reuse or recycling, and protection from depletion and degradation. There is an urgent need to protect groundwater quantity and quality from depletion and ensure its optimal utilization. However, in India (global champion of groundwater abstraction), as over 70% of surface water resources are polluted, groundwater supports >80% of water supplies and 60–90% of irrigated agriculture, although groundwater is also polluted in many parts. In much of the northwest region, there has been shallow groundwater renewal from limited modern recharge over the past decades and by interaction with lakes and rivers. In most places, low annual recharge (<5–11% of rainfall) to shallow groundwater suggests a limited renewal potential of the dynamic component.

Since the hydrogeological, social, economic, cultural, and political factors vary greatly at regional and local scales and the aggregate impact of millions of individual decisions is difficult to assess, no single template for management can be developed. Due to diminishing water supply over recent decades, efforts on better harvesting, conservation, and reuse of resources have been taking place to improve the way resources are used for industry, agriculture, and domestic purposes. However, in the context of water governance, the rules and responsibilities are divided between central and state authorities and ineffective management has resulted in conflicts among different sectors. To reduce water consumption and wasteful utilization, practical measures should include strict regulation of groundwater extraction and waste discharge into water bodies, identification of pollution sources and containment of pollution spreading from known sources, development of vulnerability maps, and conservation of floodwater in floodplain aquifers.

For rainwater harvesting to ensure sustainable groundwater supply, development and management must be based on adequate knowledge of groundwater systems, considering environmental, geographical, economic, social, and political aspects. Practical decisions should be governed by ethical values (honesty, trust, reliability, transparency, etc.), giving more importance to citizens' welfare and economic development than to private interests.

The public services reform programs focus on administrative policy and legalities but overlook governance arrangements. Policy discourse has only recently started to consider governance of water supply as a key issue. The technologies are considered as an important dimension in the design of supply systems. The widespread use of groundwater creates a vast supplement to consumption, and is important for financial, economic, and ecological sustainability of supply. However, groundwater use is mediated by economic and political power via social and technological systems, to keep tariffs very low by not increasing coverage. Thus, it is more readily available to powerful groups, leading to depletion in the overall resource and negative effects for poorer users and the hydrological system as a whole.

In this context, even though the title of this book refers specifically to rainwater harvesting and groundwater management, it intends to add to the literature considerably beyond both, extending to what water harvesting exactly means, overview of the historical evolution of water harvesting for groundwater recharge; benefits and gaps in knowledge; implementation and funding strategies; public participation; and the strengths, weaknesses, opportunities, and threats of the harvesting technologies. Although emerging threats are dynamic and diverse, efforts can be made to address the issues and also to analyze the extent to which artificial recharge approaches have been successful in finding solutions to water scarcity problems.

Chapter 1 aims to provide general information and understanding of the issues, concept, and principles of rainwater harvesting and water security; food-water-energy nexus; climate and environmental impacts (floods and droughts); and discusses the challenges in achieving groundwater security. Chapter 2 outlines mega-trends of growth in population, urbanization, agricultural and economic aspirations; increased demand for water, food, and energy; and anticipated climate change that can affect water security. Chapter 3 describes basic information on groundwater provenance, occurrence, availability, recharge, residence time, pollution characteristics; identification of recharge zones vulnerable to floods; and risk mitigation. Chapter 4 describes historical phases of water harvesting systems; artificial groundwater recharge; its local benefits and achievements so far; possible negative impacts; governance problems; and to what extent water harvesting by artificial recharge technologies can help to solve water scarcity. Chapter 5 provides international scenario and global case studies of water harvesting in selected countries with differing climates, incomes, and lifestyles; and successes and failures of water harvesting for groundwater recharge.

Chapter 6 highlights SWOT analysis of water harvesting systems; strengths, weaknesses, opportunities, and threats; and issues/limitations/questions that need attention. Chapter 7 describes challenges associated with water harvesting; land encroachment, land grabbing, water pollution, etc.; and risks associated with managed aquifer recharge. Chapter 8 covers scope of water harvesting for groundwater management strategies; background of GWM; artificial recharging methods; requirements of sustainable integrated water resources management; good governance in solving sustainability challenges; demand management tools to achieve groundwater resource conservation targets; performance of governance in transition, legal, and policy frameworks.

Chapter 9 describes guidelines to make water harvesting helpful and meaningful for GWM; demand management framework, and tools for groundwater resources managers. Chapter 10 deals with future scope; policy framework; integrating ethical principles with management; and presents proposed ways to move forward in a collaborative, co-operative way to develop policies for sustained supplies, by modifying the attitudes and behavior of all users in order to balance demand with supply and avoid costly expansions to achieve natural resources sustainability for the supply network – the costs being economic, environmental, and political.

This book is distinctive in the sense that (i) there is currently no proper international definition of rainwater harvesting for groundwater management; (ii) it is difficult to extend the limited research conducted so far to sustainable groundwater management at the river basin scale and the actual processes that enable the application of GWM; (iii) while the definitions of water management tend to focus on flood, waste, and potable water collection, treatment, and distribution, the concept embraces little on the impact of water use on the natural ecosystem, making it difficult to demonstrate or compare RWH successes across areas; and (iv) an understanding of the social, economic, and political dimensions of demand for water lags significantly behind science and engineering knowledge on supply of water. This book will provide guidance to administrative, academic, and research interests and help to raise public awareness on ethical values in the effective use of water harvesting approaches for groundwater recharge, and the risks associated with managed aquifer recharge.

Partha Sarathi Datta
New Delhi, June 2018

Acknowledgments

First and foremost, I thank the Almighty's grace for enriching me with the power to believe more and more in my passion for water and to march forward for my dreams, and for making me realize how true this gift of book writing is for me, and for inspiration and motivation to continue to improve my knowledge and progress in my professional career. I thank and appreciate my wife Shakuntala and son Ankur for always standing beside me, making me feel happy to follow my ambitions, and for having the patience with me for taking on another challenge that decreases the amount of time I can spend with them. I also thank all my co-workers during my professional career, for allowing me the freedom to manage projects and provide the necessary time and resources which gave me the opportunity to think about the theme of this book. I appreciate that they believed in me to provide the leadership and knowledge which made this book a reality.

This book would not have been possible without the continuing interest of Justin Taberham, Publications and Environment Consultant, Series Editor and Advisor for Wiley Publishing and Palgrave Macmillan/Springer Nature, from the beginning. I thank Justin Taberham; Andrew Harrison, Senior Commissioning Editor, Earth and Environmental Sciences; Athira Menon, Project Editor; Anabella Talbot, Senior Editorial Assistant, all of John Wiley & Sons Publishing, Oxford, UK, for their continuing interest, and devoting time and effort to providing excellent support and advice to bring this book to the present stage. I would like to express my gratitude to all the John Wiley & Sons publishing community who provided support and assistance in the editing, proofreading, and design and helped me to make this book complete.

Last, but definitely not least, I'd like to thank all those colleagues in the Indian central and state authorities and agencies, who have been with me over the course of 40 years during my professional career, and all those in rural areas who helped in the extensive field investigations. Although I cannot mention all their names, discussions with them have been very helpful in obtaining real grassroots knowledge and understanding of rainwater harvesting problems, perspectives, and governance issues related to groundwater assessment and management.

Partha Sarathi Datta
New Delhi, June 2018

1

Introduction

Issues in Water Harvesting and Water Security

CHAPTER MENU

1.1 Concept/Principles of Water Harvesting and Water Security

Rainwater harvesting (RWH) is not a new concept. The basic principle of water harvesting (WH) is to capture rainfall and high-intensity rainfall-induced surface runoff in one area and transfer it to another water-scarce area, thereby increasing the amount of water available in the latter. The concept of water security has its genesis in the need to shift focus to demand-side water management from supply-side, so that water is available in adequate quantity, during both normal and typical conditions. It became clear long ago that water systems should be considered as a whole, since water quantity and water quality issues on rainwater, surface water, and groundwater resources are linked. However, the term "water security" has taken a central position recently associated with other popular terms such as "food-water-energy nexus," "water hazard," "water risk," "water vulnerability," "water resilience," "sustainable water resources management," "integrated water resources management," and "adaptive water management."

The term "water security" is generally associated with other terms such as "integrated," "sustainable," and "adaptive." However, achieving water security, in practice, requires interinstitutional and interdisciplinary integration across the boundaries of many sectors, such as political, administrative, governance, biophysical, social, infrastructural, economic, and financial, most of which lie outside the direct realm of water, to reduce competition or even conflict over water resources. Development studies often consider the national scale, hydrological studies generally employ a catchment scale, and social scientific studies usually focus on the community scale. Therefore, water security at household, local, urban, rural, state, regional, country, or global level is likely to have very different meanings. WH and water security have been therefore defined in a number of ways, and can be grouped as *in situ* and *ex situ* types, and it is necessary to clearly assess the potential of RWH in reality, depending on the source, availability, and volume of water.

Water Harvesting for Groundwater Management: Issues, Perspectives, Scope, and Challenges, First Edition. Partha Sarathi Datta.

1.1.1 Source of Water and Its Availability

The primary source of water is rainfall. Globally, the exposed land surfaces receive approximately 113 000 km^3 of rainfall each year. Of this, approximately 41 000 km^3 (or 36%) is surface runoff, which replenishes rivers, streams, and lakes. The balance (64%) is evaporated via vegetation, soil, and water surfaces.

Easy access to water and its adequate availability is essential for living beings, domestic, agricultural, industrial, and economic development purposes. Although, globally, enough water exists, its supply has reduced due to growing population demand for water for domestic, agriculture, energy production, manufacturing, and healthcare purposes, and via pollution. In many countries, where water supply comes mostly from surface water bodies, pollution of rivers and lakes and the non-availability of adequate quantities of good-quality water supply has lead to increased dependence on groundwater. This has resulted in indiscriminate extraction of groundwater in excess of the natural recharge in many areas, causing substantial decline in groundwater levels and yields of many dug wells and tube wells.

The increase in groundwater abstraction, associated with population growth and increasing demands for water, food, and income, began during the first half of the twentieth century in a limited number of countries such as Italy, Mexico, Spain, and the USA. However, the periods of maximum growth were not simultaneous, and varied from country to country: 1950–1970 in the United States, 1960–1990 in India, and 1975–2000 in China. The most pronounced increase in extraction has been in those countries where current groundwater withdrawals are the highest. In the 1960s, a second phase started in parts of South and East Asia, the Middle East, and northern Africa. In the 1990s, a third phase started, including some South and South-East Asian countries such as Sri Lanka and Vietnam, and also sub-Saharan Africa. In the developed countries, stabilizing/declining trends followed a strong but variable increase in abstraction of groundwater in the earlier stages.

For example, in the USA, in 30 years (1950–1980), abstraction increased by 144%, followed by a temporary decline but a stable average during 1980–2005. In Japan, in 30 years (1965–1995), abstraction increased by 60%, before declining by 13% during 1995–1999. In several European countries also, although extraction was generally at lower levels, periods of marked growth have been observed; for example, in the UK, more than 54% of growth occurred in 25 years (1950–1975); in Denmark, more than 70% in seven years (1970–1977); in Spain, more than 15% in 10 years; and in The Netherlands, more than 12% in five years (1971–1976). In Australia, groundwater abstraction almost tripled in 30 years (1970–2000). In Germany, Belgium, France, Sweden and Canada, groundwater abstraction remained relatively stable over time. Since the beginning of the twenty-first century, almost all European countries have demonstrated stabilization or slightly declining groundwater abstraction trends.

In developing countries, such as India and China, with increasing demographic and economic growth, increases in groundwater abstraction have generally been observed from the period 1970–1980 onwards, especially where irrigation has expanded significantly. Large increases have occurred in countries of the arid and semi-arid zone where oil revenues facilitated deep groundwater (including non-renewable resources) to be withdrawn for irrigation. Total groundwater withdrawal increased by a factor of 11 in Libya during 1970–2000; by 10 in Saudi Arabia during 1975–2000; by 6 in Egypt during

1972–2000; by 3.3 in Iran during 1965–1995; and by 3.2 in Tunisia during 1977–2000. During the last 50 years, easy accessibility of groundwater and its local availability have increased the number of shallow wells. On the other hand, public affordability for sinking tube wells has increased due to technological developments in construction of deep tube wells, water abstraction devices and pumping methods, provision of free or subsidized electricity for pumping in many parts, and easy credit availability from financial institutions.

Globally, this has resulted in indiscriminate extraction of groundwater in excess of the natural recharge in many areas, causing substantial decline in groundwater levels and yields of many dug wells and tube wells, particularly during the summer. It has become difficult for resource managers to co-ordinate users of the same aquifers across wide geographic spaces. Over the last few decades, increasing groundwater depletion and degradation in India and other parts of the world, and ecologically damaging, socially intrusive, capital-intensive, and unsustainable water resource development projects implemented so far have forced people to find alternative sources of water supply. These problems have forced people to consider local RWH, conservation, and reuse for agriculture, irrigation, and other purposes. Globally, from time immemorial, people were aware that groundwater augmentation by artificial recharge and WH by collecting, storing, and conserving of local rainwater surface runoff on natural catchment areas on rocky surfaces, hill slopes, rooftops, and artificially prepared impervious/semi-pervious surfaces may provide augmentation of water supply.

1.1.2 Concept and Definition of Water Harvesting

The RWH concept employs a wide range of approaches and technologies to collect and store rainwater. WH has been defined and classified in a number of ways, and can be grouped as *in situ* and *ex situ* types, depending on the source of the collected water.

a) *In situ* RWH technologies are soil management strategies that enhance rainfall infiltration and reduce surface runoff.
b) *Ex situ* RWH collects surface runoff water from areas such as rooftops, land surfaces, steep slopes, road surfaces or rock catchments for storage in tanks. Depending on the size of the storage, *ex situ* RWH systems can be further divided into passive and active systems.
   - Passive systems (e.g. rain barrels) are small-volume (50–100 gal) systems that capture rooftop runoff without treatment. The captured water is generally not used for drinking purposes. Due to their size, passive harvesting systems are commonly used in residential applications.
   - Active systems (e.g. cisterns) are of relatively larger volume (1000–100 000 gal) and capture runoff from roofs or other suitable surfaces such as terraces and road surfaces. Active harvesting systems provide water quality treatment and can be used on a community level.

Commonly, the terms WH and RWH are synonymous and used interchangeably. The main difference is with respect to breadth and scope. Most of the definitions are similar and closely related, and WH is generally defined as: "The collection and management of high intensity rainfall induced water runoff and/or floodwater to increase water availability for domestic and agricultural use as well as ecosystem sustenance."

1.1.2.1 Why Harvest Rainwater?

Important reasons for harvesting rainwater include the following:

- For centuries, it has been the simplest indigenous technology, being practiced in India and many other parts of the world.
- To conserve surface water runoff during the monsoon.
- To reduce soil erosion.
- To arrest groundwater decline and augment the groundwater table.
- To improve water quality in aquifers.
- To inculcate a culture of water conservation.
- Self-sufficiency, less expensive, and ease of maintenance.

1.1.2.2 Aims of Water Harvesting

The aims of WH are essentially to collect surface runoff from areas of surplus and to store it in over- or underground storage, to recharge groundwater levels, deliberately reallocate water resources over time within a landscape and make it available to places where there is water shortage. This increases the availability of water by (i) impeding and trapping surface runoff, (ii) maximizing water runoff storage, and (iii) harvesting subsurface groundwater. Water harvesting makes more water available for domestic, livestock, and agricultural use by buffering and bridging drought spells and dry seasons through storage. Water harvesting captures water for domestic use, replenishes green water supplies, or increases blue water availability locally. New methods and mechanisms are being developed all over the world to conserve water as far as possible in all sectors. Our predecessors also adopted concepts of water recharge, which are still practiced today. However, with time, improvements in technology have introduced new recharge techniques to achieve better results.

1.1.2.3 Principles, Concept and Components

The basic principle of WH is to capture rainfall in one area and transfer it to another water-scarce area, thereby increasing the amount of water available in the latter. Water harvesting is seen as an integral part of sustainable land and water management. Water harvesting and runoff recycling has six basic components: a catchment or collection area, collection (harvesting) of excess rainfall, runoff conveyance system (including lifting and conveyance), water application area, efficient storage of harvested water, and optimal utilization of applied water for maximum benefits. In some cases, the components are adjacent to each other, in other cases they are connected by a conveyance system. The storage and application areas may also be the same, typically where water is concentrated in the soil for direct use by plants.

1.1.2.3.1 Catchment or Collection Area

The area where high-intensity rainfall-induced runoff is harvested. The catchment area may be a few square meters to several square kilometers. It may be a rooftop, a paved road, compacted surfaces, rocky areas or open rangelands, cultivated or uncultivated land and natural slopes.

1.1.2.3.2 Conveyance System

The system by which runoff is conveyed through gutters and pipes (in the case of rooftop WH) or overland via rills, gullies or channel flow and diverted onto cultivated fields (where water is stored in the soil) or into specifically designed storage facilities.

1.1.2.3.3 Storage Component

The place where harvested runoff water is stored until it is used by people, animals, or plants. Water may be stored in the soil profile as soil moisture, above ground (tanks, ponds or reservoirs), underground (cisterns) or as groundwater (near-surface aquifers) (Oweis et al. 2012). In places where concentrated runoff is directly diverted to fields, the application area is identical to the storage area, as plants can directly use the accumulated soil water. A great variety of storage systems hold the water until it is used either adjacent to the storage facilities or further away.

1.1.2.3.4 Application Area or Target

The area where the harvested water is put into use for domestic consumption (drinking and other household uses), livestock consumption, or agricultural use (including supplementary irrigation).

Water harvesting may occur naturally, for example in depressions, or "artificially" through human intervention. Artificial WH involves interventions to improve precipitation collection and to direct runoff to the application area. When surface runoff is very low, approaches such as smoothing or compacting the soil surface, clearing rock surfaces, surface sealing or using impermeable coverings can improve WH. Water availability by harvesting can be enhanced by the recharge of soil water and groundwater, and water storage in reservoirs, for ecosystem maintenance and industrial use, although most WH technologies are employed for domestic and agricultural use.

The catchment to application area ratio (C/A) represents the quantity of rainfall/runoff in WH systems, and is a measure of the size of the catchment compared with the size of the application area. It is generally used where runoff is stored in the soil for plant production. In the design of WH systems, this ratio is determined by considering seasonal rainfall, crop water requirement, and physical characteristics of both the catchment and the concentration area. Ideally, the catchment area (with the exception of rooftop WH) should have clay or shallow soils with low infiltration rates, susceptible to crusting and sealing, or hard surfaces with high runoff coefficients such as roads or rocky hillsides. However, in systems where runoff is stored in the soil, deep soils with high water infiltration and storage capacity are desirable in the application area.

1.1.2.4 Water Retention, Recharge, and Reuse Concept

This concept focuses on water buffering to better manage natural recharge, and to extend the chain of water use. During very high-intensity rainfall, large quantities are usually lost through floods, surface runoff, and evaporation. The unused water can be retained through the following buffering technique/strategies.

- *Groundwater recharge and storage*: this is "closed" underground storage with smaller evaporation losses than for open water storage. Water can be accessed by wells. Examples include sand dams, infiltration ponds, and spate irrigation.
- *Soil moisture conservation in the root zone*: water is stored in the root zone of the soil. Crops can use some of the water, and the remaining part percolates deeper to recharge the groundwater. Examples include grass strips, deep plowing, and conservation agriculture.
- *Closed tank storage*: in this approach, water can be stored without pollution, close to the location where it is used as drinking water. Examples include rooftop tanks, underground cisterns, and fog shields.

- *Open surface water storage*: by this method, larger volumes of water can be stored and used for agricultural and industrial purposes. Examples include storage reservoirs (small), road WH, and trapezoidal bunds.
- *Coastal reservoirs for freshwater storage*: the excess fresh water from river runoff during the wet season can be stored in reservoirs and then pumped to the land during the dry season, to meet the water demand and supply gap in coastal regions.

Each type of buffer option has its strengths and weaknesses, and local conditions usually help to determine which option to use. In general, buffering capacity increases as one moves from small to large storage, and from surface to soil or groundwater storage. Often, different types of storage complement each other in water buffering at landscape and basin level.

1.1.2.5 Water Harvesting for Groundwater Recharge

Water harvesting for groundwater recharge is most generally used for collecting and managing high-intensity rainfall-induced floodwaters and runoff to recharge groundwater artificially. The term "artificial recharge" has different connotations for different practitioners. Whether the groundwater recharge occurs under natural or artificial conditions, the same physical laws govern the recharge process. However, in artificial recharge, human effort involves adopting procedures and approaches to accelerate the natural process of recharging the aquifers through percolation of stored or flowing surface water which otherwise would not easily reach the aquifers.

In the broadest sense, artificial recharge of groundwater refers to the addition and/or infiltration of surface water to the aquifer to ensure the availability of water at a particular location at a particular time. Therefore, any man-made facility that adds water to an aquifer may be considered as artificial recharge. Artificial recharge also aims at augmenting groundwater storage at a rate exceeding that of natural conditions of replenishment, by some method of construction, spreading of water, or by artificially changing natural conditions. The process may be either planned, such as storing water in pits, tanks, etc. for deliberately feeding the aquifer, or unplanned and incidental to human activities, such as applied irrigation, leakages from pipes, etc. This is useful for reducing overdraft, conserving surface runoff, and increasing available groundwater supplies.

Rainwater in rural areas is considered fairly clean except for some dissolved gases it may pick up from the atmosphere. However, rainwater in urban areas is not free of pollution and contains atmospheric gases in dissolved form, sediments, dust, aerosols, particulates, anthropogenic gases from industrial discharge, biomass and fossil fuel burning, metallurgical processes, and other anthropogenic activities, and also biochemical processes in soil and water. The carbonates, nitrates, and sulfates in the atmosphere react with water vapor and form carbonic, nitric, and sulfuric acids, which create acid rain, which is detrimental to ecosystem and water quality. Therefore, both in rural and urban areas, surface runoff may carry pollutants from non-point sources (fertilizers, pesticides, chicken and cow manure), dissolved minerals, sediments, sewage, decaying plants, algae, bacteria, etc. Building elevated banks around ponds can reduce surface runoff, but water may be lost to evaporation if the ponds are not completely covered. Moreover, ponds may be connected to groundwater flow and may be subject to contamination by dissolved chemicals.

In RWH harvesting, high-intensity rainfall-induced floodwater and surface runoff can be harvested to recharge and replenish groundwater, depending on catchment type. Water is conserved and stored for reuse to extend crop-growing periods and/or for supplementary irrigation during dry periods in semi-arid rural areas. Groundwater extraction is done in traditional as well as unconventional ways (e.g. qanat systems, horizontal wells, etc.). The major advantages of RWH are that it is simple, efficient, cheap, replicable, and easily adaptable at small scale, and it improves water use efficiency and soil fertility, reduces soil erosion, and improves agricultural productivity (Li et al. 2000). However, the quantity of water that can be harvested depends on the frequency and intensity of rainfall, catchment characteristics, runoff amount, water infiltration rate and percolation rate to recharge the aquifers.

1.1.3 Factors Governing Selection of Suitable Structures for Artificial Recharge of Groundwater

- Quantity of non-committed surface runoff available
- Rainfall pattern
- Land use and vegetation
- Topography and terrain profile
- Soil type and soil depth
- Thickness of weathered/granular zones
- Hydrological and hydrogeological characteristics
- Socioeconomic conditions and infrastructural facilities available
- Environmental and ecological impacts of artificial recharge scheme

From time immemorial, people have been aware that traditional RWH technologies with simple, highly efficient structures for collecting and conserving local rainwater that runs off natural/man-made catchment areas can provide water for the local society and meet their needs for future use. Therefore, rainwater collection through artificial storage has been practiced in semi-arid areas. Due to its many benefits, RWH has received increased attention in recent decades and both governments and NGOs have started promoting RWH as a solution to water scarcity and water access. Consequently, a rapid expansion of rainwater catchment systems has occurred, especially in Asian and African countries facing water scarcity. RWH is mostly a local intervention and a decentralized method of gaining access to drinking water. However, due to very limited and sketchy knowledge about actual field-level performance in specific locations for augmentation of groundwater resources, debates have been ongoing for or against the large-scale acceptance of these technologies.

For instance, in India, despite over two decades of publicized activities and critical acclaim for these practices, with investments by international agencies for bringing "traditional knowledge" into development practice and encouragement from those sympathetic to grassroots environmentalism, concern about their vulnerability still exists. These technologies have also faced wide-ranging criticisms for ignoring scientific knowledge (and thus effectiveness) and power relations (linked to issues of equity) (Mosse 2003; Chhotray 2007). Funding agencies experience limitations in implementation at local scale, due to minimum scientific quantitative analysis of real-time data about the status of aquifer water quality and quantity, in order to understand the balances between water budget zones, yields of various crops grown, etc. in order to predict the impact of decisions on aquifer water quality and quantity – a common decision support system for all stakeholders.

Moreover, due to heterogeneity and the interconnectivity of groundwater aquifers across space and with surface water sources, recharge by one group or community may impact water availability for other neighboring or downstream groups. So far, impacts have been reported mostly with respect to changes in water availability, increased groundwater table, revival of flow of rivulets, and yields of crops, relying mainly on qualitative analysis based on secondary information collected from farmers/beneficiaries. However, limited/no scientific quantitative analysis has been done based on baseline data and chronological documentation to actually quantify the extent of soil erosion prevented, rise in the water table, increase in the time of water flow, or yields of the various crops grown at basin/watershed level.

In many countries, due to pollution of surface water bodies and non-availability of adequate quantities of good-quality water, there has been increasing dependence on groundwater. Unplanned and indiscriminate extraction of groundwater has also risen due to an increase in the sinking of tube wells driven by affordable technological developments in construction of deep tube wells, water abstraction devices and pumping methods, provision of subsidized or free electricity for pumping in many areas, and easy credit from financial institutions. This has resulted in imbalance between water withdrawal and natural recharge in many areas, causing substantial decline in groundwater levels and yields of many dug wells and tube wells. Therefore, it is necessary to determine the strengths, weaknesses, opportunities, and threats of these WH technologies, as well as benefits and gaps in knowledge, implementation and funding strategies, public participation, etc., and to assess the ability of these technologies to overcome the water stress situation in any meaningful sense.

Against this background of RWH and groundwater issues, an analysis of the food-water-energy nexus is discussed in the next section.

1.2 Food-Water-Energy Nexus

Water is essential for agricultural production, and energy and food security. Water, energy, and food create an interrelated nexus of resources. Food production is the largest consumer of water globally, accounting for 30% of global energy demand, as petroleum-based fertilizers and transportation are critical to the food production supply chain. Water is used to generate 8% of world energy. Energy production creates pollution from fossil fuels, damaging agricultural land and water. The nexus is context/regional specific, with variations in water resources due to geography, population, economic growth, demand, energy mix, and climate. It is therefore important to understand the regional challenges and devise context-specific solutions to address the nexus. The regional variability of water supply and the associated costs of water supply infrastructure for energy needs can significantly affect energy planning, especially in a water-scarce country. For policy makers, the food-water-energy nexus is a balancing act between competing issues of human security.

In the "nexus" approach, efforts are made to use new policies and tools to improve resource values in order to achieve higher resource use efficiency, which cannot be achieved in isolation. The nexus is a conceptual framework that considers the interconnectedness of water, food and energy securities, and seeks to integrate and develop collective solutions and strategies that promote positive synergies among these sectors, mitigating the tradeoffs and focusing on the interlinkages among the three resources. According to UN reports, effective institutional arrangements, which can facilitate policy integration in developed and developing countries, still suffer from considerable uncertainty due to differences in environments. Low- and middle-income countries tend to use most of their water for meeting basic needs such as food production, whereas the share of water used for domestic and industrial consumption increases exponentially in high-income countries.

Worldwide, over 80% of available water (particularly groundwater) is used for agricultural activities and food crop production. In combination with water, other factors such as high-yielding hybrid seeds and availability of supplement nutrients like fertilizers, etc. are responsible for increasing food production. Water consumed daily through food intake is much higher than water used for drinking. The food consumed by one person daily needs 2000–5000 L water to produce. Hence, growing water scarcity is now one of the most significant challenges for food security. Therefore, along with rapid population growth, the interconnection of land, water, and energy use for water supply, agriculture and food production, and associated water pollution is also increasing. The complexity and interdependency of water, energy, and food systems are growing with increasing human, social, and economic development. Projections on climate change are making it imperative that in order to manage water and energy availability, allocation, production, consumption, and security, the food-water-energy nexus is integrated into all development plans.

One problem is that most people have limited concern about higher consumption of water, energy, and food, or the wasteful use of water and energy, and little knowledge about balancing water-for-food, water-for-energy, energy-for-food, and energy-for-water. Little is known about people's attitude and behavior patterns on consumption and use of water, energy, and food. In reality, policy makers often ignore the food-water-energy nexus, despite the fact that water and energy are needed to grow food, energy is needed to pump groundwater, and water is needed to generate energy, even in solar farms. For example, pumping groundwater provides over 50% of irrigation in South Asia, and the aspirations of the growing South Asian population increase demands on energy and food. But in the current global situation, when droughts, floods, and storms are more frequent and severe, organizations should be increasing consultation with other sectors, within and between different countries, and examining all the implications.

For those working across sectors, scope exists for more rigorous research and advocacy to bridge key knowledge gaps, share knowledge, and explore more efficient approaches to addressing the interdependencies of the water, energy, and food sectors. Water, energy, and food exist in a complex web of vulnerability to each other, due to the following reasons.

- Water is required throughout the energy and food sector, in power generation of most forms of turbine-generated electricity, either directly (hydropower, geothermal) or indirectly (steam to turn turbines, cooling). Fresh water is required for food production, energy extraction and production, refining and processing, transportation and storage.

- Energy, primarily mechanical or electrical energy, and sometimes in the form of human/animal power, is required to extract, move, treat, deliver, use and dispose of water, and produce food. Energy supplies are harnessed to pump water from deeper groundwater reserves and to divert rivers over large distances.

1.2.1 Water Use Per Unit of Energy

Water use per unit of energy is generally determined in terms of "withdrawal" and "consumption." Withdrawal is the volume of water removed from a source and consumption is the volume of water that is not returned to the source after withdrawal; that is, it is evaporated or transported to another location. "Discharge" is the volume of withdrawn water that is often degraded (physically or chemically) by use, and is returned to the source, affecting its water quality. Since all types of energy and food production require large amounts of water, and water withdrawals are always greater than or equal to consumption, hence, when water availability declines and there is not enough to satisfy demands, it is important to first limit/reduce consumption of water at energy and food production facilities.

Therefore, as populations increase and droughts occur, amount of water usage is of great concern for electricity-generating systems. Oil presently accounts for only 10% of water consumption in primary energy production. Energy from coal production is currently less than oil but is likely to rise over the next 30–40 years. The mining and refining of coal requires water at various stages. Overall, the production of coal accounts for about 1% of total water consumption in energy production. Almost 90% of fresh water is used for the production of biomass, which accounts for hardly 10% of total primary energy production. In 40 years, the share of fresh water used to produce biomass will decrease to less than 80%, while the share of biomass in total primary energy production will reduce to <5%.

1.2.2 Water Consumption to Generate Electricity

Water consumption to generate electricity is projected to more than double over the next four decades. The current annual electricity generation per capita average of 2.9 MWh (ranging from 0.6 MWh in Africa to 12.0 MWh in North America) may almost double in 2050 to an average of 5.7 MWh, ranging from 2.0 MWh in Africa to 17.3 MWh in North America. The highest increases may occur in Latin America, where per capita electricity generation is likely to be four times higher than currently, followed by Africa and Asia, where it may be almost triple. In Europe, per capita electricity generation may double while in North America, it may increase by only 50%. Although worldwide electricity generation per capita may almost double, the amount of water consumed to generate electricity is likely to stay at the same level due to technology advancements, or increase only slightly on a per capita basis in Africa, Europe and North America, whereas in Asia and Latin America, per capita water consumption to generate electricity will almost double.

As traditional energy resources decrease, the attraction of unconventional sources (e.g. oil sands, oil shales, deep gas shales) increases. However, many of these require large amounts of water, further stressing current and projected systems. For energy production, policy makers must consider available water supplies, understand the driving forces, relationships and water and energy cycles for efficient and sustainable use of these technologies, and search for ways to maximize the supply of one while minimizing the use of the other. The efficiency of power plants can result in decreased water use to generate the same amounts of electricity.

The increasingly integrated trade in water for energy production and conversion requires a new paradigm of co-operation among regional and national governments, and businesses. The water-energy nexus can become a vicious cycle, as lack of technology and poor management or inefficiencies in use in one region can affect the sustainability of another. A more plausible approach may be to preserve the present level in developed nations while reducing primary energy needs and continuously improving energy intensity, possibly with present technology and large investments.

1.2.2.1 Minimizing Water Use in Energy Production

According to the US National Renewable Energy Laboratory, electricity production from fossil fuels and nuclear energy requires 190 000 million gallons of water per day, accounting for 39% of all freshwater withdrawals in the nation. In many regions of the country, as much water is used for turning on lights and running electrical appliances in homes as in taking showers and watering lawns. Of this, 72% is used for fossil fuel-related energy production, and coal accounts for 52% of US energy generation. Each kWh of energy generated by coal requires 25 gal water. According to the United States Geological Survey, thermoelectric power generation consumes only 3.3% of fresh water, with over 80% for irrigation. According the US Department of Energy, average freshwater usage (gal/MW-h) of different power sources is: nuclear (400–720); coal (390); natural gas (140); hydroelectric (1430); solar thermal (1060); geothermal (2900); biomass (390); photovoltaic (30); wind (1). All thermal cycle plants (nuclear, coal, natural gas, solar thermal) require large amounts of water for steam, cooling, and condensing. The amount of water needed reduces with increase in boiler temperatures so coal, which burns at very high temperatures, is more efficient and requires less water use.

Both water quantity and quality issues are also important. For instance, electricity companies prefer to use cooling water from a lake or river or a cooling pond, instead of a cooling tower, if environmental feasibility and cost are acceptable. Pumping cooling water through the heat exchangers of the plant can reduce the energy costs of cooling towers. However, discharge of the waste heat may increase the temperature of the water source. Power plants using natural water bodies for cooling purposes have to be designed to prevent intake of organisms into the cooling cycle. Complications may arise if organisms adaptable to the warmer plant water are injured when the plant shuts down in cold weather. Thermal cycle plants require water for cooling, but it need not be fresh water. A power plant located on the coast can use sea water without using cooling towers. Moreover, discharge water temperatures would have less effect on the environment. If dry cooling systems are used, significant water from the water table need not be used.

A large amount of energy is also needed to extract, treat, and supply potable water, and further to collect, treat, and dispose of waste water. Technological advances can help to reduce energy requirements for water processing in water treatment, specifically in the desalination of salt water, and in the treatment of contaminated water and waste water for reuse. The key technologies for water treatment are reverse osmosis, ion exchange, and ultrafiltration. New approaches to reduce the energy footprint of water treatment systems involve using waste energy throughout the treatment process.

Innovative desalination technologies that use low-grade or waste heat instead of electricity have the potential to substantially reduce energy inputs, leading to an environmentally benign process and lower operation costs.

These water treatment technologies help power plants utilize available water supplies efficiently. Specialized ion exchange resins also help uranium mining operations use less water and generate less waste and enable the production of high-purity uranium used in nuclear power applications, helping to meet the increasing global demand for energy. However, demands for water and energy are increasing faster than technology can advance. The demand for water in highly populated parts of the world is high, and the resources needed to provide it are not always adequate. For instance, in China, both economic growth and physical scarcity drive water scarcity. China's annual water deficit is ~40 bcm in normal years, and 50% of its cities are facing some degree of water shortage.

Water shortages are a challenge for India as well. With 85% of the available water currently being used for agriculture, a further depletion in water supply could jeopardize agriculture and food security. Canal water and groundwater are the major sources of irrigation. Groundwater irrigation has expanded rapidly in recent decades, and forms a major part of the water withdrawals in many river basins. At present, the groundwater-irrigated area constitutes more than 60% of the total irrigated area. During recent decades, there has been practically no scope to further increase the canal water-irrigated area, and in the last 10 years or so, there has been a decline in the canal water-irrigated area.

In India, the near self-sufficiency in food production allows export of some of it with hard-to-estimate externalities including groundwater depletion and salinization, as well as degradation of soil health and the environment. India is the world's third largest exporter of groundwater through its grains export. Going by per capita availability, the parts of India from which most grains are exported are seriously water scarce. Average groundwater consumption to grow 1 kg wheat is 812 L, 1 kg rice takes 200 L (because it is far more dependent on surface water) and maize 72 L. In the 2016–2017 financial year, India is estimated to have exported 3 00 000 tonnes of wheat, 10.7 million tonnes of rice, and 700 000 tonnes of maize. India is also the world's largest extractor of groundwater; around 75 km^3 groundwater was extracted in India in 2010.

Water is usually inefficiently used in the food production chain. Decisions on site selection, technology, and suppliers are frequently made without considering the impacts of the operation on the availability and quality of water resources, especially when water is not a limiting factor by either quantity or price. When considering grain exports, it should be kept in mind that water embedded in the grains is the actual water exported, and this is much smaller than the total water used to grow crops because a small fraction of the total amount of water used in growing crops is recoverable, and the rest seeps down to the groundwater. A much larger amount of water is lost due to evaporation, and this is key to understanding how water-intensive crops affect groundwater in a region. An additional related factor for India is state-level disparities in groundwater depletion; dry regions of Gujarat and Karnataka are exporting water to wet regions of the country to satiate the thirst of those who can afford bottled water and soft drinks.

Within the power sector in India, coal-fired plants account for around 95% of total water withdrawals, the rest being split between gas-fired and nuclear power stations.

The cooling technology used, together with the overall efficiency of the power4plants, determines the amount of freshwater withdrawal from local sources, and the amount withdrawn but not returned to the local water basin (water consumption). Hence, for future industrial growth in India, availability of water has to be ensured, with water requirements for manufacturing and power generation competing with safe drinking water supply and agricultural purposes. In India, along the coast, new coal-fired power plants primarily rely on imported coal and use sea water as a cooling medium, giving them a cost advantage for transport and limiting their water stress.

A mismatch between water supply and water demand is also common in the developed world. Under an average economic growth scenario and without efficiency gains, global water requirements will grow from 4500 billion cubic meters today to nearly 7000 billion cubic meters – a 50% increase in only 20 years. By 2030, some analysts predict that available water supplies will satisfy only 60% of demand. There is no doubt that, for the survival of humanity, water and energy are and will always be inextricably linked in virtually everything from growing and processing food to industrial processes that require energy.

1.3 Climate Forcing Environmental Impacts (Floods and Droughts)

Several studies have reported increases over recent decades in the frequency of heavy to extreme hydrological events. Climate change is expected to accelerate global hydrological cycles, resulting in increased precipitation and reduced evapotranspiration, thereby increasing river discharge on a global scale. Several regions were projected to have increases in both flood frequency and drought frequency. Such regions show a decrease in the number of precipitation days, but an increase in days with heavy rain. Several regions show shifts in the flood season from springtime snowmelt to the summer period of heavy precipitation. Temperature and precipitation data have been used to estimate the Palmer Drought Severity Index for the USA, for projecting future drought (mainly as a shortage of precipitation). However, low river flows (droughts) must also be considered and measured, because many regions are irrigated with river water. The low river flows may not be estimated simply from soil moisture, rainfall or precipitation minus evapotranspiration. Drought frequency has been projected to increase globally, with a decrease or no significant changes in regions such as northern high latitudes, eastern Australia, and eastern Eurasia.

However, changes in flood and drought are not explained simply by changes in annual precipitation, heavy precipitation, or differences between precipitation and evapotranspiration ($P - E$). In contrast, river basins characterized by decreases in $P - E$ in the twenty-first century show increases in drought, but an increase in drought does not necessarily result in a decrease in $P - E$. The changes in precipitation patterns may cause an increase in the number of drought days even though precipitation, discharge, and $P - E$ increase. Comparison of flood and drought frequencies estimated from daily discharge is therefore important in predicting future discharge extremes.

According to the UN Millennium Development Goals 2007 Report, the primary contributor to climate change is the increasing level of atmospheric CO_2 from burning of fossil fuels. Carbon dioxide emissions continue to rise, as evidenced by increasing concentrations in the atmosphere. In South-East Asia and northern Africa, emissions more than doubled during 1990–2004. On a per capita basis, developed regions continue to emit far more CO_2 than developing regions. Among developing regions, western Asia is the highest per capita emitter. Sub-Saharan Africa accounts for per capita less than 10% of the CO_2 produced by an average person in the developed world. The depleting groundwater tables may force users to lower the depth of pumps in the wells, increasing the energy requirement and fossil fuel use for pumping water and thereby increasing the emission of CO_2 to the atmosphere.

Changes in flood extremes using monthly averaged river discharge data for both gauge observations and global climate model (GCM) simulations over 29 river basins larger than 200 000 km^2 in area suggest that the risk of great floods increased during the twentieth century, mainly over the northern high latitudes, and that this increase may continue. Future extreme daily precipitation projections have also been estimated based on various climate change simulations. The frequency of floods was projected to increase over many regions, except North America and central to western Eurasia. However, these studies are insufficient for future projections of extremes in global river discharge in GCM simulations, because extremes in river discharge are strongly affected by the spatial distribution of the precipitation intensity within a basin, and due to the limitation of spatial resolution in GCMs for the accurate estimation of extremes.

The intensification in extremes is therefore not easily determined from observations or current GCMs without considering anthropogenic water usage. Future global warming projections using other GCMs, regional climate models (RCMs), and ensemble models with higher spatial resolution are therefore expected to provide multimodel projection of discharge extremes.

One type of flood problem or risk is runoff which exceeds the capacity of the drainage channels – rivers and creeks in the rural environment, and storm water channels in the urban environment. Sudden floods can cause sudden morphing of existing streams into torrents, totally altering the landscape, rocks, foothills, channels, and stream heights. Water pollution and food security can ensue, due to difficulty in delivering safe water and food supplies to flood-affected areas. Risk mitigation plans often involve emergency resources, where groundwater is readily accessible locally. Groundwater from deep aquifers or even non-renewable "fossil" water bodies needs to be tested for adequate yields.

Flood risk mitigation for safe water supply requires a search for groundwater resources, with long residence times, proven safe and protected by the physical environment, along with the necessary infrastructure for exploitation, with timely replacement of vulnerable water supply systems. For identification of low-vulnerability aquifers, sustainable management of groundwater resources, groundwater renewability and rock–water interactions must be considered based on systematic hydrogeological investigations including monitoring and mapping of groundwater, integrated with geochemical and other methods. Identification of such aquifers requires risk analysis with respect to their occurrence, accessibility, quality, vulnerability, and resistance to the impact of disasters.

Shallow uppermost unconfined aquifers mainly occur in unconsolidated fluvial and glacial deposits overlain by permeable unsaturated zones of low thickness (<10 m), characterized by young groundwater and single flow system and interface with surface water. Deeper unconfined aquifers in consolidated rocks (particularly sandstones) of regional extent, overlain by permeable unsaturated zones of variable thickness, consist usually of a number of laterally and vertically interconnected groundwater flow systems of dual porosity and permeability. Karstic aquifers occur in carbonate rocks with groundwater flow in conduits, large open fissures, and openings along bedding plates, typically with high groundwater flow velocities (100 m day^{-1}) and secondary permeability; springs are important phenomena of groundwater karstic regimes. Coastal aquifers under natural conditions consist of sea water intrusion, controlled by tidal fluctuations, stream flow changes, gradient and volume of groundwater flow towards the seashore, and geological environment. Groundwater pumping significantly influences the groundwater–sea water interface. Recharge areas of all type of aquifers need to be reliably identified/delineated.

Some controversial questions always remain. Will there be no risk if there is no vulnerability? What risk is negligible? What risk is tolerable? What risk is intolerable? In view of the aforementioned uncertainties, since an extreme flood- or drought-induced disaster can dramatically affect the impact and recovery time before, during, and immediately thereafter, in the context of RWH for groundwater management, the behavioral adaptation dynamics of society, individuals, groups, and government agencies need to be seriously taken into consideration for more reliable risk characterization and assessment of the effectiveness of risk management strategies, policies, and investments. However, existing risk assessment methods rarely include this.

As global energy consumption continues to expand (20% increase since 1990), there has been progress in the development and use of renewable resources, such as hydropower and biofuels. The development of more modern renewable resources, which have no negative impact on people's health or the environment, has increased 10-fold over recent decades. However, these newer technologies, including those relying on wind, solar, wave, and geothermal energy, still account for only 0.5% of total energy consumption.

Large regions are at risk from the effects of both climate change and acidification. All major cities in the world suffer multifaceted pervasive problems. In Eastern Europe, air quality is considered the most serious environmental problem. Acid rain and transboundary air pollution, once problems only in Europe and parts of North America, are now becoming apparent in parts of the Asia-Pacific region as well as Latin America. Despite co-ordinated action worldwide, non-compliance and growth in illegal trade in ozone-depleting substances are emerging problems, with continued damage to the ozone layer occurring faster than expected. The coming decade is predicted to be the most vulnerable. All regions express concern over global warming, but special emphasis is placed by developing countries on the need for adaptive mechanisms to cope with accompanying climate variability and sea level change. The rapidly rising demand for energy for economic development can only aggravate these problems.

All regions experience problems related to groundwater or surface water or both. With the global population forecast to grow over the next 40 years, especially in Africa and Asia, water demand will increase, making it even scarcer than it already is in many regions of the world. The efficient development and management of water resources is a priority concern in West Asia, Africa, and Asia and the Pacific. In the Middle East, China, and India also, which already suffer from water stress or water scarcity, water stress will increase further, making the whole region very vulnerable. With these trends, it is important to understand the impact an increased use of renewable energy and improved energy efficiency could have in terms of water savings. Increased competition for water calls for more integrated approaches to the use of natural resources.

In Europe and North America, the protection of water resources from contamination, acidification, and eutrophication is highest on the agenda. Other global priorities are the equitable distribution of water among riparian countries sharing international river basins, non-point sources of pollution, and the impacts of major dams and diversion projects. More than 60% of all water is withdrawn in Asia, more than 10% by Europe and North America, around 5% by Latin America and the Caribbean, and even less in Africa. The proportion of the population using safe sources of drinking water in the developing world increased from 71% in 1990 to 79% in 2002. The most impressive gains were made in southern Asia. However, at present, billions of people are still using water from unsafe sources, and water-borne diseases represent the single largest cause of human sickness and death worldwide. More than one-third of the world's population is without a safe water supply.

About 60% of the world population lives within 100 km of the coast, and more than 3 billion people rely in some manner on coastal and marine habitats for food, building sites, transportation, recreation, and waste disposal. Around 33% of the world's coastal regions are at high risk of degradation, particularly pollution and infrastructure development from land-based sources. European coasts are most affected, with some 80% at risk, followed by Asia and the Pacific, with 70% of the coast at risk. In Latin America, some 50% of the mangroves are affected by forestry and aquaculture activities. Oil spills are threats in West Asia and the Caribbean, while tourism industry infrastructure development puts stress on natural coastal areas around the world.

1.4 Water Security

The definition of water security proposed by UN Water to serve as a starting point for dialogue in the UN system is: "The capacity of a population to safeguard sustainable access to adequate quantities of acceptable quality water for sustaining livelihoods, human well-being, and socio-economic development, for ensuring protection against water-borne pollution and water-related disasters, and for preserving ecosystems in a climate of peace and political stability." According to Wikipedia, water security has been defined as "the reliable availability of an acceptable quantity and quality of water for health, livelihoods and production, coupled with an acceptable level of water-related risks." Water security also means addressing environmental protection and the negative effects of poor management. It is also concerned with ending fragmented responsibility for water and integrating water resources management across all sectors: finance, planning, agriculture, energy, tourism, industry, education, and health. A water-secure world reduces poverty, advances education, and increases living standards.

However, stakeholders from different disciplinary backgrounds seem to have different perceptions of the term "water security." For instance, engineering studies generally focus on water supply and demand, protection against floods, droughts, and contamination. Water resources studies focus on water scarcity, water supply, and demand management. Environmental studies usually focus on environment, water availability in terms of quality and quantity, and on impacts of hydrological variability. Policy studies focus on linkages of food, climate, energy, economy and human security, protection against water-related hazards, and sustainable development of water resources. Public health studies focus on water supply security and access to safe water, and prevention of water pollution in distribution systems. Political and governance perspectives focus on institutional division of responsibilities, power structures, equity issues, and water planning conflicts. Legal perspectives focus on water rights and ownership. Economic perspectives focus on the efficiency of water resource use, the economics of water demand and supply, water pricing and market mechanisms, cost–benefit analysis of flood risk protection, and water quality conservation.

It is therefore important that in framing policies to make water available for food, governances in different countries ensure that water is also available for energy production and conversion. Without energy to supply the water needed for all uses, there can be no production of food or the economies of modern food processing. Asia, with large geographical area and population, faces the largest challenge for water supply in general and for water use in energy production. While designing multiple resource management frameworks, the increasingly integrated shared resources and trade mechanisms require a new paradigm for the sustainable global availability and use of resources in the coming decades and beyond, instead of each country or region looking out only for itself. To effectively address these complexities, policies and frameworks need to integrate knowledge, understanding, specialization, and cross-disciplinary approaches to managing natural resources simultaneously.

References

Chhotray, V. (2007). The anti-politics machine in India: depoliticisation through local institution building for participatory watershed development. *Journal of Development Studies* 43 (6): 1037–1056.

Li, F., Cook, S., Geballe, G.T., and Burch, W.R. Jr. (2000). Rainwater harvesting agriculture: an integrated system for water management on rain-fed land in China's semi-arid areas. *Ambio: A Journal of the Human Environment* 29: 477–483.

Mosse, D. (2003). *The Rule of Water*. New Delhi: Oxford University Press.

Oweis, T.Y., Prinz, D., and Hachum, A.Y. (2012). *Water Harvesting for Agriculture in the Dry Areas*. London: CRC Press.

Further Reading

Asano, T. (1985). *Artificial Recharge of Groundwater*. Boston: Butterworth.

Bhalge, P. and Bhavsar, C. (2007). Water management in arid and semi arid zone: traditional wisdom. International History Seminar on Irrigation and Drainage, Tehran, pp. 423–428.

Borthakur, S. (2008). Traditional rainwater harvesting techniques and its applicability. *Indian Journal of Traditional Knowledge* 8 (4): 525–530.

Dhiman, S.C. and Gupta, S. (2011). *Rainwater Harvesting and Artificial Recharge*. New Delhi: Central Ground Water Board, Ministry of Water Resources.

Huisman, L. and Olsthoorn, T.N. (1983). *Artificial Groundwater Recharge*. Massachusetts: Pitman Publishing.

2

Mega-Trends that Impact Water Security

CHAPTER MENU

2.1 Global Population Growth and Water Availability

Rapid growth in population, increasing industrialization, and emerging economies need ever-increasing quantities of water. All over the world, in the next two decades or so, water demand to support population growth and economic growth will outstrip usable supply by 40%. Against this background, water scarcity refers to historical use of fresh-water resources to sustain life, community, and industry, associated with a range of other human interests, and the fact that the proportion of water resources that continue to be usable is in fact shrinking with increasing demand for water for various purposes.

In five decades (2000–2050), the world population is projected to grow dramatically, from 6 billion in 2000 (1 billion in the developed world and 5 billion in the developing world) to 9 billion in 2050. One hundred million out of the 3 billion global increase in population will be in the developed world, while the developing countries will increase by 2.9 billion. According to available projections, by 2030, 2 billion middle-class people will be in Asia (1 billion in China by 2050). Africa will grow to 2 billion people by 2050. By the year 2050, the average per capita income in the US, Europe, Africa China, and India will also increase. By the end of the century, more than half the population is likely to live in urban areas, possibly increasing to 60% by 2020. Europe, Latin America, and North America may have more than 80% of their population living in urban areas. On the other hand, driven by growth in global population and rapid transfer of modern technology to developing countries, by 2050, these countries are likely to have 65% of the global economy, while the developed world may have 35% of the world economy (Figure 2.1).

All around the world, people tend to want to settle down and plan for safe and peaceful future. However, over 1.3 billion people, mostly in the rural areas of developing countries, are suffering from land degradation. Along with water, land will continue to be used for urban needs, agriculture, forestry, recreation, and other purposes. Therefore, growth in population-induced demand for food and water is projected to double by

Water Harvesting for Groundwater Management: Issues, Perspectives, Scope, and Challenges, First Edition. Partha Sarathi Datta.

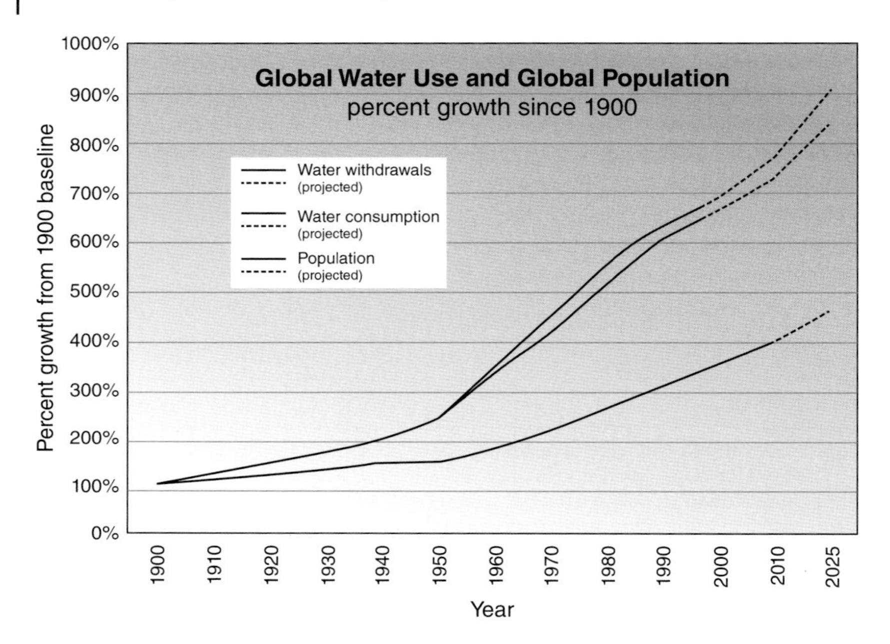

Figure 2.1 Global water use and global population growth. *Source:* United Nations Environment Program.

2050, but crop yields are projected to fall on drought-affected and degraded land. While the global demand for water is rising steadily, water supply has substantially declined. There is always strong competition for water and land, usually determined by economic factors. Going by the land area used for crop production, in societies with strong social commitments, partitioning of land for its multiple uses is determined by the population-supporting capacity, societal values, and economic considerations.

Although poverty and the growing global population are often targeted as responsible for much of the degradation of the world's resources, other factors, such as the inefficient use of resources, waste generation, pollution from industry, and wasteful consumption patterns, are equally significant and important matters of immediate concern. The trends and changes outlined above may have significant impacts on all sectors of the global transfer in economy, production, services, and technology. Due to these effects, rapidly increasing demand has caused a rise in the cost of resources, particularly since 2003.

With soaring population, urbanization, and industrialization, fresh water, food, and energy are becoming increasingly scarce. Both potable and industrial water use is limited, and some supplies are inadequate. Two-thirds of all the water consumed is used for irrigating crops. The need for water in the developed world is high. In the developing world, the challenge is making the limited amount of available fresh water clean enough to drink. Under an average economic growth scenario and without efficiency gains, global water requirements will grow from 4500 billion cubic meters today to nearly 7000 billion cubic meters by around 2037 – a 50% increase in only 20 years. Some predictions suggest that, by 2030, available water supplies will satisfy only 60% of demand. According to the World Health Organization, around 1.1 billion (17%) of the world population does not have access to safe water for drinking. The world is also facing significant challenges for energy production. Clearly, clean water and energy are the two intimately intertwined critical needs, strongly related to income demand and development for a productive healthy society.

When the land does not provide sustenance or the economic opportunity to benefit from increasing global demand and wider economic growth, people tend to migrate on a temporary basis, in order to survive. Over the next few decades, worldwide, around 135 million people will be at risk of being permanently vulnerable due to land degradation and desertification. Most importantly, more and more people are migrating to coastal areas. In Australia, where in the 1950s only 70% of people lived in the coastal areas, this percentage has now increased to 92%. Around the world in 1950, only 30% of the world's total population lived in coastal regions, but presently about 70% live there. More and more mega-cities are developing in the coastal zones such as New York, Tokyo, Shanghai, Mumbai, Beijing, Tianjin, etc. which leads to a worsening water deficit.

Although yearly rainfall is not the only factor that determines whether or not a country encounters water scarcity, variability in rainfall and losses due to evaporation and runoff also contribute to the amount of water available. Areas that experience moderate to high amounts of rainfall are less likely to be affected by physical water scarcity. For instance, the annual rainfall in some parts of Africa is comparable to Europe and North America. However, higher evaporation losses in the African continent result in a lower percentage of rainfall contributing to renewable water resources. The high variability of rainfall, especially in Africa and Asia, makes it even more difficult to predict how much rainfall can be used for various purposes. South Asia, mainly India, experiences monsoon-type rainfall from June to September. However, throughout the rest of the year, rainfall in India is moderate to low. Northern Africa experiences heavy rainfall in the summer months, while southern and equatorial Africa experience monsoon-type rainfall in spring and autumn. The uneven distribution often leads to a cycle of floods in the rainy seasons and droughts in the dry season.

For human survival, water, land, and energy are always inextricably linked to everything from developmental activities, food growing and processing to industrial processes that require energy. Let us take an example. The aspiration for economic growth and a higher standard of living is directly linked to consumption of fossil fuel but reserves of fossil fuel have a lifetime of only decades. The demand for energy (mostly as fossil fuel) during 1970–2010 increased most rapidly in the least developed countries, from an average 30 billion barrels of oil equivalent/year in 1970 to 60–70 billion barrels of oil equivalent/year in 2010, due to aspirations to achieve the living standards of industrial countries. The BP Statistical Review of World Energy (2016) data also suggest a different kind of shift in the distribution of oil supplies. China, Saudi Arabia, and India show significant increases in oil consumption (from 100% in 2000 to 180–250% in 2015). At the same time, many of the developed countries (e.g. US, Japan, Italy) found their oil consumption shrinking (from 100% in 2000 to 70–90% in 2015).

The increasing global demand for energy and other natural resources, such as water, coal, oil, and gas, is already reflected in their rising costs. However, burning coal, oil, and natural gas for electricity and heat production causes about 25% of global greenhouse gas emissions. Worldwide, it is projected that in around two decades, wind and solar power sources alone could account for 48% of installed electricity capacity, from 12% presently, and 34% of electricity output from the present 5%. According to reported projections, global emissions from electricity are likely to peak in 2026, as governments and companies shift focus from coal towards wind and solar power; after this peak, emissions will decline by 1% every year until 2040. This is in contrast to the International Energy Agency's forecast, which expects emissions to rise steadily for decades to come. In view of these trends, it is important to understand what impact increased use of renewable energy and improved energy efficiency could have in terms of water savings. Increased competition for water calls for more integrated approaches use of natural resources.

The BP Statistical Review of World Energy (2016) data from 1999 to 2015 suggest that the quantities of energy goods production tend to increase and decrease in a similar manner, as the costs of oil, coal, and natural gas tend to rise and fall together. One reason for this may be that as the world economy moves swiftly forward (higher wages, more building activity, more debt), demand tends to be high for many different types of materials at the same time. When the economy slowed during 2011–2015, prices (in US $) of oil, coal, and natural gas fell by almost 50% at the same time (with a steep fall during 2014–2015) in Japan, Germany, UK, and US. Associated with this, inflation also tended to fall.

2.2 Increased Demand for Water, Food, and Energy

All regions experience problems related to either groundwater or surface water or both. With the global population forecast to grow over the next 40 years, especially in Africa and Asia, water demand will increase, making it even scarcer than it already is in many regions of the world. Human consumption could deplete groundwater in parts of India, southern Europe, and the US in the coming decades. According to studies of the world's groundwater levels, aquifers in the Upper Ganges Basin area of India, southern Spain, and Italy could be depleted between 2040 and 2060. In the US, aquifers in California's Central Valley, Tulare Basin, and southern San Joaquin Valley could be depleted within the 2030s. Aquifers in the southern High Plains, which supply groundwater to parts of Texas, Oklahoma, and New Mexico, could reach their limits between the 2050s and 2070s.

It is projected that by 2050, about 1.8 billion people could live in areas where groundwater levels are fully or nearly depleted because of excessive pumping of groundwater for drinking and agriculture. While many aquifers remain productive, economically exploitable groundwater is already unattainable or will become so in the near future, especially in intensively irrigated areas in the drier regions of the world. A new study finds that heavily irrigated regions in drier climates, such as the US High Plains, the Indus and Ganges basins, and portions of Argentina and Australia, face the greatest threat of depletion. Although this study estimates the limits of global groundwater on a regional scale, due to lack of complete data about aquifer structure and storage capacity it is difficult to say exactly how much groundwater remains in individual aquifers. The efficient development and management of water resources is a priority concern in western Asia, Africa, and Asia and the Pacific. In the Middle East, China, and India also, which already suffer either from water stress or water scarcity, water stress will increase further, making the whole region very vulnerable.

In Europe and North America, the protection of water resources from contamination, acidification, and eutrophication is high on the agenda. Other global priorities are the equitable distribution of water among riparian countries sharing international river basins, non-point sources of pollution, and the impacts of major dams and diversion projects. More than 60% of all water is withdrawn in Asia, more than 10% is withdrawn in Europe and North America, around 5% is withdrawn in Latin America and the Caribbean, and even less in Africa. The proportion of population using safe sources of drinking water in the developing world increased from 71% in 1990 to 79% in 2002. The most impressive gains were made in southern Asia but at present, billions of people are still using water from unsafe sources, and water-borne diseases represent the single largest cause of human sickness and death worldwide. More than one-third of the world's population is without a safe water supply.

Food production in India has gone up from about 50 million tonnes per year in the 1950s to over 250 million tonnes per year in recent decades, with a corresponding fourfold increase in yield in terms of kilograms of total grains per acre. The Kharif (summer) season used to produce more than twice the tonnage of grain as the Rabi (winter) season during the nascent years of independent India but now the two seasons are nearly equal in grain production. The combination of many factors such as water, high-yielding hybrid seeds, availability of supplement nutrients like fertilizers, etc. is responsible for improving food grain production. However, the parallel trends of food grain production and area under irrigation shown in Figure 2.2 clearly illustrate that water resource has contributed the most to this increase.

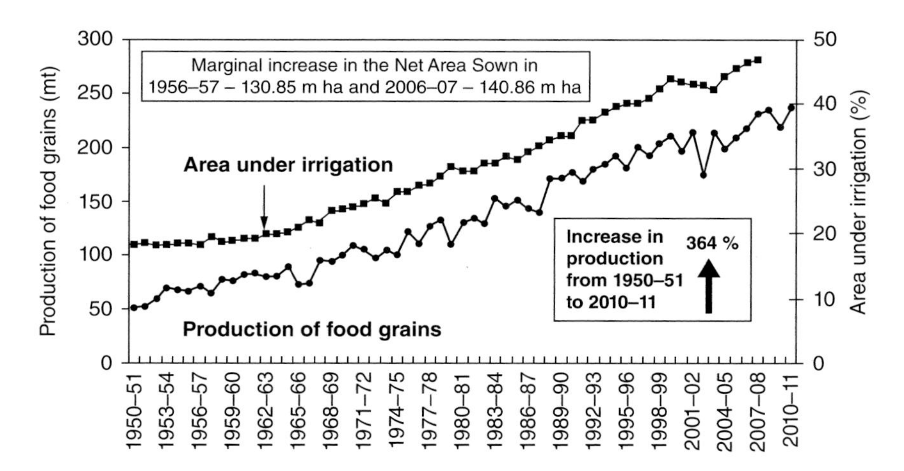

Figure 2.2 Trends of food grain production and area under irrigation in India. *Source:* Directorate of Economics and Statistics, Department of Agriculture and Cooperation, India, 2008–2009, 2010–2011.

In India, canal water and groundwater are the major sources of irrigation. During recent decades, there has been practically no scope to further increase the canal water-irrigated area, and there has been a declining trend in the canal water-irrigated area in the last decade or so. In contrast to this, groundwater irrigation has expanded rapidly in the last few decades, and forms the major part of water withdrawal in many river basins. The contribution from groundwater to the total irrigation area has increased substantially, from 5 978 000 ha (29%) in 1950–51 to 35 372 000 ha (59%) in 2005–06. At present, groundwater-irrigated areas constitute more than 60% of the total irrigated area. However, over the last decade or so, even the groundwater-irrigated area has shown a relatively slower rate of growth compared to the exponentially expanding situation from 1950 to 2000. This may possibly be due to declining groundwater levels induced by indiscriminate withdrawal, thereby increasing the costs, energy requirement, and fuel requirement for pumping, and also due to pollution of groundwater in many areas. Wheat accounts for 35% of total groundwater depletion, and rice accounts for about 25%, while fodder, cotton, and sugarcane make up the rest.

Water is essential for agricultural production and food security. The daily water consumption through food intake is much greater than water used for drinking. The food consumed by one person daily needs 2000–5000 L of water to produce. Hence, growing water scarcity is now one of the most significant challenges for sustainable development. It is therefore important that governments, in framing policies to make water available for food production, ensure that it is also available for energy production and conversion. Without energy to supply the water needed for all uses, there can be no production of food or the economies of modern food processing. Asia, with its large geographical area and population, faces the largest challenge for water supply in general and for water in energy. The increasingly integrated shared resources and trade require a new paradigm, instead of each country or region looking out only for itself.

Water is usually inefficiently used in the food production chain. Decisions on site selection, technology, and suppliers are frequently made without considering the impacts of the operation on the availability and quality of water resources, especially when water is not a limiting factor by either quantity or price. When considering grain exports, it should be kept in mind that water embedded in the grains is the actual water exported, and this is much smaller than the total water used to grow crops because a small fraction of the total amount of water used in growing crops is recoverable, and the rest seeps down to the groundwater. A much larger amount of water is lost due to evaporation, and this is key to understanding how water-intensive crops affect groundwater in a region. An additional related factor for India is state-level disparities in groundwater depletion; dry regions of Gujarat and Karnataka are exporting water to wet regions of the country to satiate the thirst of those who can afford bottled water and soft drinks.

According to new modeling of the world's groundwater levels, unplanned indiscriminate overexploitation and consumption of groundwater could deplete groundwater aquifers in the Upper Ganges Basin area, and many other parts of India, the US, southern Europe, Spain, and Italy could be depleted between 2040 and 2060. In the US, aquifers in California's Central Valley, Tulare Basin, and southern San Joaquin Valley could be depleted within the 2030s. Aquifers in the southern High Plains, which supply groundwater to parts of Texas, Oklahoma, and New Mexico, could reach their limits between the 2050s and 2070s. By 2050, as many as 1.8 billion people may live in areas where groundwater levels are fully or nearly depleted because of excessive pumping for drinking and agriculture.

While many aquifers remain productive, economically exploitable groundwater is already unattainable or will become so in the near future, especially in intensively irrigated areas in the drier regions of the world. A new study finds that heavily irrigated regions in drier climates, such as the US High Plains, the Indus and Ganges basins, and portions of Argentina and Australia, face the greatest threat of depletion. Although this study estimates the limits of global groundwater on a regional scale, scientists still lack complete data about aquifer structure and storage capacity to say exactly how much groundwater remains in individual aquifers, and how fast the aquifers are depleting, or how long this resource will last before drying up of wells or rivers occurs.

The rate at which the Earth's groundwater reservoirs are being depleted is constantly increasing. Annual groundwater depletion during the first decade of this century was twice as high as it was between 1960 and 2000. India, the USA, Iran, Saudi Arabia, and China have the highest rates of groundwater depletion. About 15% of global groundwater consumption is not sustainable, meaning that it comes from non-renewable groundwater resources. On the Arabian Peninsula, in Libya, Egypt, Mali, and Mozambique, and in Mongolia, over 30% of groundwater consumption is from non-renewable groundwater. Based on improved groundwater consumption data, the estimate of global groundwater depletion is 113 000 million cubic meters per year for the period from 2000 to 2009, which is lower than previous, widely varying estimates.

The increased use of groundwater for irrigation also results in a rise in sea levels. According to one estimate, sea-level rise due to groundwater depletion was 0.31 mm per year during the period from 2000 to 2009. This corresponds to roughly one-tenth of the total sea-level rise.

2.3 Anticipated Climate Change

2.3.1 Some Historical Facts

In the 1890s, the Swedish chemist Svante Arrhenius (1859–1927) created the very first climate model and calculated the extent to which the world would warm if humans doubled or tripled the amount of carbon in the atmosphere; amazingly, the numbers were close to what the most recent global climate models (on powerful supercomputers) still find today. A Serbian concrete expert, Milutin Milankovic (1879–1958), discovered that ice ages and the warm interglacial periods are initiated by changes in the shape of the Earth's orbit around the sun and the tilt of its axis of rotation. Over time, these cycles cause the great continental ice sheets to expand and retreat. However, the warming after the last ice age peaked between 4000 and 8000 years ago. Today, according to natural cycles, the Earth should be gradually and slowly cooling, leading to the next ice age, but due to all the coal, oil, and gas burned since the beginning of the Industrial Revolution, this process has been arrested. It is well known that mining coal

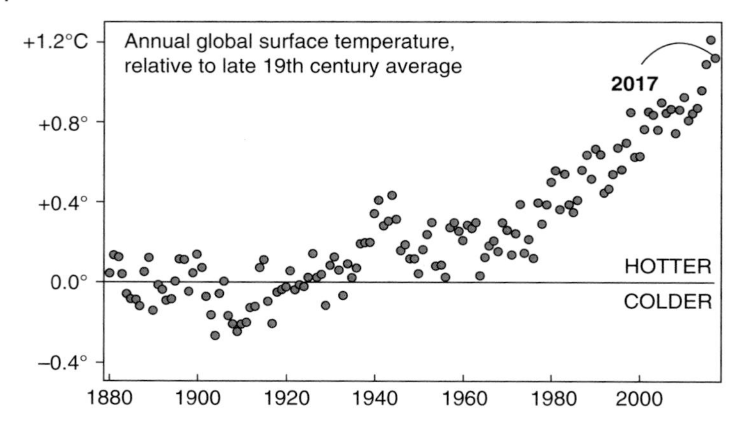

Figure 2.3 The overall trend of the temperature analyzed by NASA and NOAA.

and burning fossil fuels produces heat-trapping gases. For over 120 years, it has been known that if more CO_2 becomes artificially accumulated in the atmosphere, then it will trap more heat and the Earth will be warmer, which could cause deleterious effects from the point of view of human survival.

Globally, it is a matter of great concern that a period of climate change has been brought about by continually increasing concentrations of greenhouse gases, particularly carbon dioxide, in the atmosphere since the 1950s. The increase in CO_2 level is ascribed to use of the primary fossil fuel energy sources coal, oil, and natural gas for industrial and other development purposes. In Asia and the Pacific, 100% increase in energy use was projected for 1990–2010, and in Latin America, 50–77% energy growth was projected. It is expected that for the near future, coal, oil, and natural gas will continue to be the primary energy sources. Projections up to 2020 suggest that about 90% of Canada's greenhouse gas emissions (around 40–200 megatonnes of CO_2 equivalent) will result from fossil fuels used in manufacturing, motor vehicle production, industrial electricity generation, non-energy purposes, residential, and industrial heating. The overall trend of the temperature data (Figure 2.3) analyzed by both NASA and NOAA suggests that the Earth's climate is warming. The continuation of this phenomenon may significantly alter global and local climate, including temperature and precipitation.

2.3.2 Projections on Climate and Past Evidence

The processes controlling the exchange of energy and water vapor between the ocean and atmosphere govern the large-scale drivers of regional climate change and determine the dynamics of ocean and atmosphere. However, their predictive ability is poor. The Fifth Assessment Report of the Intergovernmental Panel on Climate Change (IPCC), published in 2013 and 2014, estimated that the global mean surface temperature has increased $0.6 \pm 0.2\,°C$ since 1861, and predicted a 2–4 °C increase over the next 100 years. Since 1976, every year including 2013 had an average global temperature above the long-term average. On average, the Earth's atmosphere warmed by about 0.17 °C per decade from 1970 to 1998, but by about 0.04 °C per decade from 1998 to

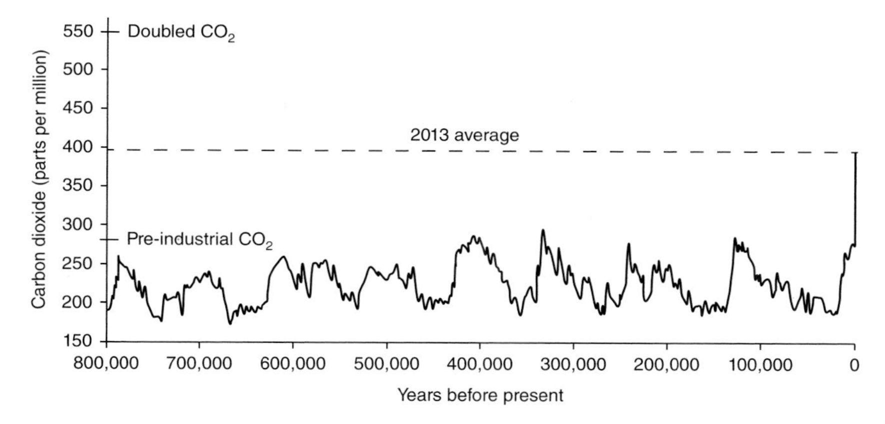

Figure 2.4 Atmospheric carbon dioxide concentrations in parts per million for the past 800 000 years, based on EPICA (ice core) data, with the 2013 annual average concentration of 396.48 ppm (dashed line) appended. The peaks and valleys in carbon dioxide levels follow the coming and going of ice ages (low levels) and warmer interglacials (higher levels). *Source:* NOAA Climate.gov, based on EPICA Dome C data from the NOAA NCDC Paleoclimatology Program.

2012. Over 1976–2013, the temperature warmed at an average of 0.50 °F (0.28 °C) per decade over land, and 0.20 °F (0.11 °C) per decade over the ocean.

Preliminary data for 2013 showed that the annual average carbon dioxide concentration was around 396 parts per million (ppm). In recent years, carbon dioxide concentrations have been growing at a rate of 2 –2.5 ppm $year^{-1}$. At those rates, it would take 60–80 years to double the preindustrial level of 275 ppm. However, the rate of increase over the past half century has not been steady but has been accelerating by about 0.5 ppm $year^{-1}$ each decade. If this rate of growth continues in the future, then doubled preindustrial carbon dioxide concentrations will be reached in about 50 years. Scientists say that doubling preindustrial carbon dioxide levels will likely cause global average surface temperature to rise by between 1.5° and 4.5 °C (2.7°–8.1 °F) compared to preindustrial temperatures (Figure 2.4). (Current concentrations are about 1.4 times preindustrial levels.) The full process could take hundreds or even more than a thousand years.

During the industrial era, along with greenhouse gases, aerosol particle pollution was also added to the atmosphere. The particles reflect sunlight and interact with clouds in ways that cool climate, partially offsetting warming. The likely range of temperature published in the Fifth Assessment Report from the IPCC is the same as in their first report, published in 1990. The global economic and other impacts are "difficult to estimate" and depend on a large number of "disputable" assumptions. Moreover, many estimates do not allow for catastrophic changes and factors. Many estimates of economic losses are based on the outputs of integrated assessment models (IAMs), which attempt to integrate the key elements of biophysical and economic systems. For instance, there is evidence that temperature increases of between 1.5 °C and 2 °C would lead to differing extents of sea-level rise and extreme weather events.

It has to be kept in mind that the exact relationship between a certain carbon dioxide level and a specific temperature depends on other factors, including atmospheric dust and volcanic eruptions, amount of forests and other vegetation, and variations in incoming sunlight, which have also varied over time, sometimes warming and sometimes cooling the climate. For example, anticipated projections on temperature and rainfall in Delhi, India, by the IPCC (IPCC 1996, 2001, 2007), based on 16 global climate models (GCMs) (discrete grid) and three population-induced emission scenarios (self-reliant, independent nation; rapid economy growth and balanced energy; integrated, ecofriendly, and rapid economy growth), indicate that the differences in temperature are likely to emerge only after 2030, and there is no clear evidence of significant change in rainfall (Figure 2.5).

Observational records (1901–2010) and reconstructed premonsoon anomalies (1725–2000) using tree ring (*Tectona grandis*, Pinus, Picea, Cedrus, Abies, etc.) chronology data (Pant 2003) clearly suggest a decreasing trend or insignificant change in temperature and rainfall in India; the climate during the past 250 years was not significantly different from the present (Dash et al. 2007), possibly due to meso-scale influences in the lower atmosphere (Figure 2.6).

Although greenhouse gas emissions have increased steadily upward, for 15 years global mean temperatures have mysteriously failed to keep pace. The "hiatus" in global warming is often misconstrued. It is not that warming has stalled, but rather that the rate of increase in warming has not been as high as predicted. The reasons for this are not clear, and reflect uncertainty about climate sensitivity to greenhouse gas emissions. The World Meteorological Organization (WMO) report notes that 2010 was not only the warmest year on record, but also the wettest globally. Floods were the most frequent extreme events during 2001–2010, and this trend appears to be continuing; so far in 2016–2017, we have seen significant floods in eastern Australia, central Europe, India, and Canada. However, such anthropogenic effects cannot be seen in the short term, but manifest as changing patterns associated with long-term natural climate and an increase in the frequency and intensity of extreme weather events. In addition, it is wrong to attribute individual events to climate change without in-depth understanding of physical causes and the statistical probability of these

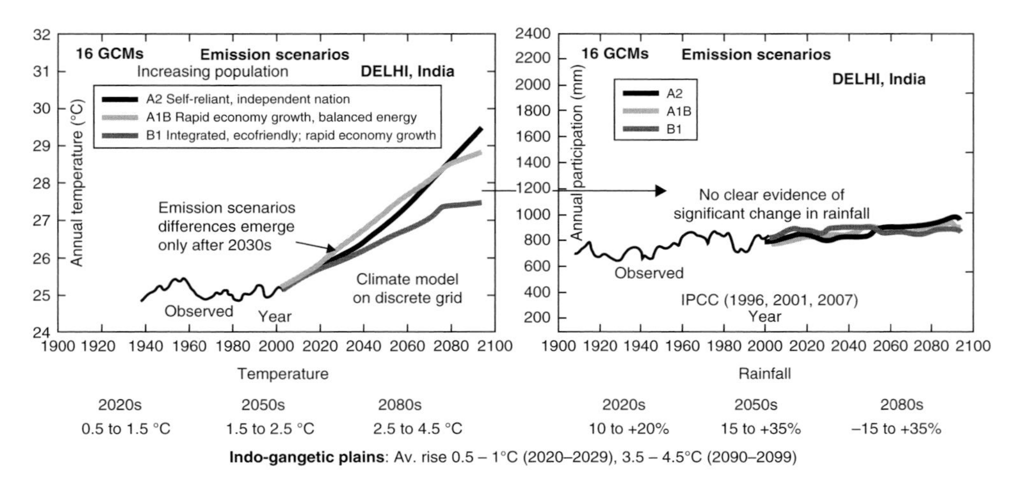

Figure 2.5 Anticipated projections on temperature and rainfall in India. *Source:* International Panel on Climate Change (1996, 2001, 2007).

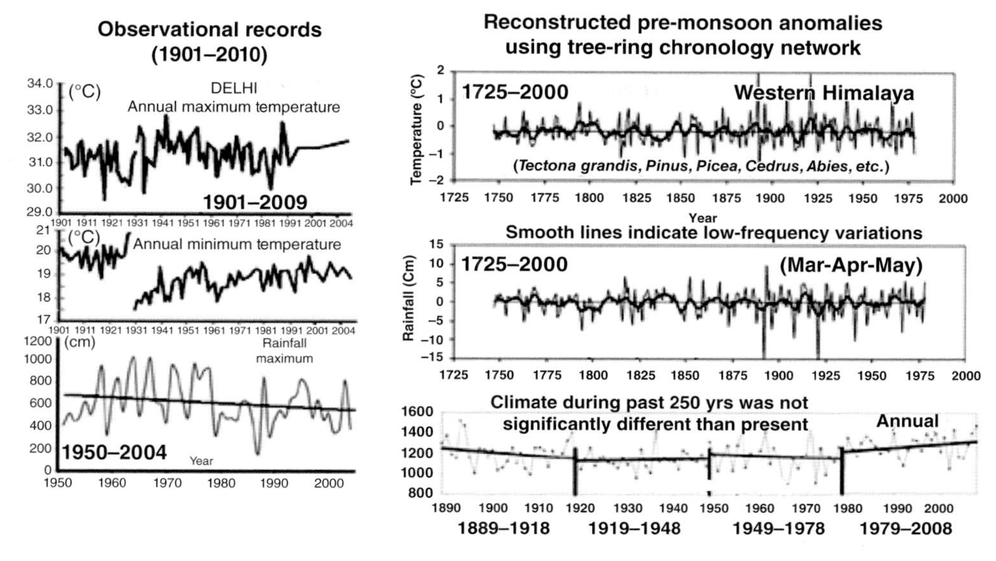

Figure 2.6 The observational records (1901–2010) and reconstructed premonsoon anomalies (1975–2000) using tree ring chronology data (Pant 2003) suggest a decreasing trend or insignificant change in temperature and rainfall in India (Dash et al. 2007).

events in relation to long-term natural variability. In error, the general public often misunderstands weather change as climate change.

Models based on observational records of temperature and rainfall, and isotopic studies on paleo-climate conditions provide no clear evidence on climate change in India during the last 1000 years BP. Temperature and rainfall show declining trends or no significant change in last 50 years (Datta 2013), and also suggest that in the early quaternary north-west India was well watered. Deep aquifers were created during the Last Glacial Maximum (LGM) (30 000–12 000 years BP) pluvial climate, and interaction with lakes and rivers recharged shallow aquifers. Going by the IPCC projections, in India, the anticipated increase in mean annual temperature from global warming may decrease winter rainfall, increase summer rainfall and surface runoff, reduce groundwater recharge where surface runoff declines, and cause more accumulation of wastes.

Despite debate on the magnitude of climate-induced warming, evidence of weather fluctuations and influence exists in some parts of the world. However, uncertainties and knowledge gaps in climate models and limited understanding of the factors controlling climate change make it difficult to assess the probable impacts in tropical regions. In this context, analysis of the available data on rainfall and temperature in India, ensemble models based on $\delta^{18}O$ and δD signatures in rainfall and groundwater and surface water, ^{14}C-ages of groundwater and lake sediments, reconstructed paleo-climate and long-term processes, and the annual mean temperature trends indicate both warming/cooling in different parts of India in the past and during 1901–2010. Neither the GCM nor the observational record suggests any significant climate change during the last 1200 years BP. One of the fundamental challenges has always been modeling uniformity across systems, which are known to have wide variability in both space and time.

Unfortunately, climate model forecasts, although known to be imperfect and rather "virtual reality," are used in decision-making processes as "deterministic" quantities. Thus, decision makers assume that the values from models predict what will actually occur and directly compare to real quantities such as thresholds, set on the basis of measured (real) quantities, without clearly distinguishing between validation uncertainty and predictive uncertainty. Climate science is still in its infancy, with limited knowledge about how the climate is changing and whether humans are responsible, and a lot more information is needed before coming to a consensus. The present philosophy seems to be to recognize, admit, and allow that political and public as well as private decisions should be taken with a subjective attitude (bias). This is particularly true for decisions under conditions of uncertainty, which are difficult to grasp and understand, decisions whose consequences are uncertain. And yet, most researchers strive to provide decision makers and society with tools that expose the meanings of uncertainty and the consequences of making decisions under uncertainty, so that decisions can be made with "open eyes" and result in minimum future regret.

Climate change with increasing temperatures can have profound effects on the hydrological cycle through more evaporation and precipitation, evapotranspiration, and loss of soil moisture. However, the extra precipitation may be unequally distributed around the globe, with some parts receiving significantly reduced precipitation or major alterations in the timing of wet and dry seasons. Consequently, these changes can influence precipitation amounts, timings and intensity rates, and indirectly impact the flux and storage of water in surface and subsurface reservoirs (i.e. lakes, soil moisture, groundwater). While climate change affects surface water resources directly, the relationship between changing climate variables and groundwater is more complicated and poorly understood. Groundwater resources are related to climate through the direct interaction with surface water resources, such as lakes and rivers, and indirectly through the recharge process. The direct effect on groundwater resources depends upon the change in the volume and distribution of groundwater recharge.

Therefore, quantifying the impact of climate on groundwater resources requires not only reliable forecasting of changes in the major climatic variables, but also accurate estimation of groundwater recharge. Attempts have been made to calculate the rate of recharge by using carbon-14 isotopes and other modeling techniques. This has been possible for aquifers that are recharged from short distances and after short durations. However, recharge that takes place from long distances and after decades or centuries has been problematic to calculate with accuracy, making estimation of the impacts of climate change difficult. The medium through which recharge takes place is often poorly understood and very heterogeneous, again challenging recharge modeling. In general, there is a need to intensify research on modeling techniques, aquifer characteristics, recharge rates, and sea water intrusion, as well as monitoring of groundwater abstractions. This research will provide a sound basis for assessment of the impacts of climate change and sea-level rise on recharge and groundwater resources.

In addition, there may be other associated impacts, such as sea water intrusion, water quality deterioration, potable water shortage, etc. The greater variability in rainfall may cause more frequent and prolonged high or low groundwater levels and saline intrusion in coastal aquifers due to sea-level rise, depending on local groundwater gradients. Shallow coastal aquifers are very sensitive to change and risk, particularly in many low-lying islands. For small island states, such as some Caribbean islands, sea water intrusion into fresh-water aquifers has been observed due to overpumping of aquifers. Decrease in precipitation coupled with sea-level rise would not only cause a diminution of the harvestable volume of water; it also would reduce the size of the fresh-water lenses. The combined fresh water from all the rivers discharging into the ocean globally is on average 42 000 km^3 $year^{-1}$. The challenge for the development of this runoff is to keep river water from mixing with sea water, as the saline water tends to mix with the fresh water in estuaries under the influence of waves and tides. Coastal reservoirs, fresh-water reservoirs in the sea, have emerged as a technology to store the runoff lost to the sea, with barriers or embankments separating the fresh water from the sea water.

Groundwater is the major source of water across much of the world, particularly in rural areas in arid and semi-arid regions, but there has been very little research on the potential effects of climate change. A change in the amount of effective rainfall will alter recharge, but so will a change in the duration of the recharge season. Increased winter rainfall, as projected under most scenarios for mid-latitudes, generally is likely to result in increased groundwater recharge. Shallow unconfined aquifers along floodplains, which are most common in semi-arid and arid environments, are recharged by seasonal stream-flows and can be depleted directly by evaporation. Changes in recharge therefore will be determined by changes in the duration of flow of these streams, which may locally increase or decrease, and the permeability of the underlying beds, but increased evaporative demands would tend to lead to lower groundwater storage. A thick layer of sands substantially reduces the impact of evaporation.

Aside from the influence of climate, recharge of aquifers is very much dependent on the characteristics of the aquifer media and the properties of the overlying soils. Several approaches can be used to estimate recharge based on surface water, unsaturated zone, and groundwater data. Among these approaches, numerical modeling is the only tool that can predict recharge. Modeling is also extremely useful for identifying the relative importance of different controls on recharge, provided that the model realistically accounts for all the processes involved. However, the accuracy of recharge estimates depends largely on the availability of high-quality hydro-geological and climatic data. Determining the potential impact of climate change on groundwater resources, in particular, is difficult due to the complexity of the recharge process, and the variation of recharge within and between different climatic zones.

Some undeniable facts are that climate has two stable modes – hot and cold – and in the Indian region the current "hot" state has been stable for the past 10 000 years. During the Neogene era (23–6 My), the climate in South-East Asia, of humid tropics and subtropics, was not substantially different, but the tropics and northern extra-tropics had a distinctly different response during the 100 000 year glacial-interglacial cycle, and ~19 000–3000-year precessional cycles of monsoons (Chiang 2009). While many current climate models strongly predict an anomalously wetter Central Asia, well-validated proxy reconstructions indicate the opposite, particularly prior to the late twentieth century (Anchukaitis et al. 2010). Despite differences in the climate models, linkage of the East Asian and Indian monsoons in the region is reported (Cai et al. 2006). Due to uncertainty in prediction of future rainfall and land use, the general scenarios based on GCMs and regional climate models (RCMs) make it difficult to assess the impacts of climate change on groundwater.

A number of GCMs and RCMs exists, which project a wide range of climate change outputs; assessment mechanisms result in model differences associated with the carbon cycle and cloud movement and explore the models' ability to predict climate on short-term decadal time scales. There is a need to downscale these models on a basin scale and couple them with relevant components of the hydrological cycle. Output from these coupled models, such as quantification of groundwater recharge, will help in suggesting appropriate adaptation strategies to combat the impact of climate change. Researchers across a range of disciplines must work together to help decision makers in the public, private, and non-profit sectors to rise to these challenges. Economists, in particular, need more help from scientists and engineers to devise models that provide better guidance about what will happen if we succeed or fail. However, if such discussions are viewed as a positive sign, this interest perhaps could lead to a broader dialogue among the many managers and practitioners dealing with climate impact: stochastic climatologists, scientists, geologists, engineers, and climate modelers. To make management more responsive, gaps in the science of climate change prediction need to be bridged based on investigations specific to the system in each area.

References

Anchukaitis, K.J., Buckley, B.M., Cook, E.R. et al. (2010). Influence of volcanic eruptions on the climate of the Asian monsoon region. *Geophysical Research Letters* 37: L22703.

BP Statistical Review of World Energy (2016). Available at: www.bp.com/content/dam/bp/pdf/energy-economics/statistical-review-2016/bp-statistical-review-of-world-energy-2016-full-report.pdf.

Cai, Y., An, Z., Cheng, H. et al. (2006). High-resolution absolute-dated Indian Monsoon record between 53 and 36 ka from Xiaobailong Cave, southwestern China. *Geology* 34: 621–624.

Chiang, J.C.H. (2009). The tropics in paleoclimate. *Annual Review of Earth and Planetary Sciences* 37: 263–297.

Dash, S.K., Jenamani, R.K., Kalsi, S.R., and Panda, S.K. (2007). Some evidence of climate change in twentieth-century India. *Climatic Change* 85: 299–321.

Datta P.S. (2013). Ensemble models on palaeoclimate to predict India's groundwater challenge. *Acque Sotterranee – Italian Journal of Groundwater*. 2(3). Available at: file:///C:/Users/Owner/Downloads/111-221-1-SM.pdf.

IPCC. 2nd, 3rd and 4th Assessment Reports (1996, 2001, 2007). Available at: www.ipcc.ch.

Pant, G.B. (2003). Long-term climate variability and change over monsoon Asia. *Journal of the Indian Geophysical Union* 7 (3): 125–134.

Further Reading

Agarwal, D.P., Datta, P.S., Zahid, H. et al. (1980). Paleoclimate, stratigraphy and prehistory in north and west Rajasthan. *Journal of Earth System Science* 89 (1): 51–66.

Bindoff, N. L., P. A. Stott, K. M. AchutaRao, et al. (2013) Detection and attribution of climate change: from global to regional. In: *Climate Change 2013: The Physical Science Basis. Contribution of Working Group I to the Fifth Assessment Report of the Intergovernmental Panel on Climate Change* Stocker, T. F., D. Qin, G.-K. Plattner, et al. (eds). Cambridge University Press, Cambridge, USA.

Collins, M., R. Knutti, J. M. Arblaster et al. (2013) Long-term climate change: projections, commitments and irreversibility. In: *Climate Change 2013: The Physical Science Basis. Contribution of Working Group I to the Fifth Assessment Report of the Intergovernmental Panel on Climate Change* Stocker, T. F., D. Qin, G.-K. Plattner, et al. (eds). Cambridge University Press, Cambridge, USA.

Döll, P., Lehner, B., and Kaspar, F. (2002). Global modeling of groundwater recharge. In: *Proceedings of Third International Conference on Water Resources and the Environment Research*, vol. I (ed. G.H. Schmitz), 27–31. Germany: Technical University of Dresden.

Döll, P., Müller Schmied, H., Schuh, C. et al. (2014). Global-scale assessment of groundwater depletion and related groundwater abstractions: combining hydrological modelling with information from well observations and GRACE satellites. *Water Resources Research* 50: 5698–5720.

Hallegatte, S. and Mach, K.J. (2016). Make climate-change assessments more relevant. *Nature* 534: 613–615.

Hayhoe, K. (2016). *Climate Science: It's a Lot Older Than You Think.* Climate Science Center, Texas Tech University Available at: http://blog.ucsusa.org/katharine-hayhoe/climate-science-its-a-lot-older-than-you-think.

Ichiyanagi K., Yoshimura K. and Yamanaka M. D. (2011). Estimating the origin of rainwater using stable isotopes in Sumatra Island, Indonesia. Available at: file:///C:/Users/Owner/Downloads/56-JSD-M156.pdf.

IPCC (2014a). *Impacts, Adaptation, and Vulnerability. Contribution of Working Group II to the Fifth Assessment Report of the Intergovernmental Panel on Climate Change.* Cambridge: Cambridge University Press.

IPCC (2014b). *Mitigation of Climate Change. Contribution of Working Group III to the Fifth Assessment Report of the Intergovernmental Panel on Climate Change.* Cambridge: Cambridge University Press.

Kosaka, Y. and Xie, S.-P. (2013). Recent global-warming hiatus tied to equatorial Pacific surface cooling. *Nature* 501: 403–407.

Lüthi, D., Le Floch, M., Bereiter, B. et al. (2008). High-resolution carbon dioxide concentration record 650,000–800,000 years before present. *Nature* 453: 379–382.

Ojiambo, B.S., Poreda, R.J., and Lyons, W.B. (2001). Ground water/surface water interactions in Lake Naivasha, Kenya, using delta O-18, delta D, and H-3/He-3 age-dating. *Ground Water* 39 (4): 526–533.

Phys Org. (2016) Groundwater resources around the world could be depleted by 2050s. Available at: https://phys.org/news/2016–12-groundwater-resources-world-depleted-2050s.html.

Singh, G., Joshi, R.D., Chopra, S.K., and Singh, A.B. (1974). Late quaternary history of vegetation and climate of the Rajasthan desert, India. *Philosophical Transactions of the Royal Society of London* 267 (889): 467–501.

Stern, N. (2016). Current climate models are grossly misleading. *Nature* 530: 407–409.

Todd, D.K. (1980). *Groundwater Hydrology*, 2. New York: Wiley.

Wogan, T. (2016). Earth's climate may not warm as quickly as expected, suggest new cloud studies. *Climate Earth* Available at: www.sciencemag.org/news/2016/05/earth-s-climate-may-not-warm-quickly-expected-suggest-new-cloud-studies?utm_campaign=news_weekly_2016-05-27&et_rid=34846921&et_cid=518691.

3

Groundwater Occurrence, Availability, and Rechargeability

CHAPTER MENU

3.1 Groundwater Provenance, Recharge, Residence Time, and Pollution Characteristics

In order to make rainwater harvesting effective for groundwater management, the public have to be given information about groundwater occurrence, availability, and rechargeability. Groundwater is the primary source of drinking water for about half the world's population, as well as the sole or largely dominant source of household water in rural areas and many large cities. Groundwater is also critical to food production, providing more than 40% of irrigation water worldwide. Groundwater is essential, and its importance is increasing, as in coming decades the world may need to double global food production to feed 9 billion or more people. New crop varieties, improved agricultural practices, reduced food waste, and leaner diets can all help meet this goal through groundwater. It is also the best insurance against droughts. However, many aquifers all over the world are under stress and face an uncertain future. There are many challenges of getting people to recognize the seriousness of groundwater problems, because people find it hard to appreciate a problem relating to a resource which they can't see, feel, and experience. The public's inability to see groundwater contributes to its oversight in water policy and management, as well as in agriculture, energy, environment, land use planning, and urban and rural development.

Every groundwater situation is unique in its spectrum of challenges. At one end are the aquifers in geological and climatic settings where groundwater pumping can be indefinitely maintained, because recharge during wet periods to some extent balances withdrawals during dry periods. However, many of these aquifers are shallow and are particularly vulnerable to contamination, which is a paramount concern of protecting groundwater quality. At the other extreme is the "fossil water" from aquifers having no or very limited present-day recharge. Examples include the southern High Plains in the USA and many aquifers in North Africa and the Middle East. For these systems, critical issues center on the quantity of groundwater that is economically recoverable, the rates of depletion, and

Water Harvesting for Groundwater Management: Issues, Perspectives, Scope, and Challenges,
First Edition. Partha Sarathi Datta.

management approaches to make society restrict excessive use and groundwater dependence. Many groundwater aquifer systems fall between these two extremes. These aquifers are naturally replenished, but are vulnerable to excessive groundwater withdrawals.

These issues are rarely addressed in any serious way. The effects of pumping on groundwater depletion, land subsidence, water quality, surface water, and ecosystems are at a tipping point for many of the world's most critical aquifers. Often, only a small part of the total groundwater in storage can be used without significant effects on water quality, surface water, ecosystems, land subsidence, etc. It is in these most critical groundwater aquifer systems that the concept of groundwater sustainability arises, and is among the hardest of goals to achieve. This goal can possibly be achieved by a tradeoff between present-day use and the future impacts of that use. Solving these problems and dealing with the consequences of the impacts at a societal level involves people's emotional values and views on the environment, current and intergenerational equity, and local customs, so that the impacts are acceptable.

Sustainability of groundwater resources requires management thinking on a different time scale. While groundwater planning horizons are often 5–20 years, the rivers, lakes, wetlands, and springs respond to groundwater pumping in many decades. Although some groundwater planning horizons look at 50 or even 100 years, the longer time frames may result in deliberate depletion of the resource. Groundwater laws and policies also need to be updated to reflect the dynamic interconnection of surface water and groundwater. While safeguarding groundwater is a global challenge, proper solutions to most problems exist at the aquifer, watershed, or local level. Strict enforcement of laws and related measures is often required to achieve necessary changes and accountability, but top-down management usually results in resistance among stakeholders. If the entrenched groundwater users are not actively involved in the decision-making process, there is virtually no possibility of achieving a solution.

Successful long-term management of groundwater resources requires ongoing data collection and sustained funding by government agencies with sufficient resources and expertise. Data collected over a period of decades are needed to monitor and assess the impacts of aquifer development and long-term trends. Unfortunately, except in some parts of a few developed countries, political and financial support for even the most basic data collection is usually a low priority, not only for basic environmental safeguards but also for resource monitoring and assessment. By definition, the world's most pressing environmental problems require co-operation and collective action involving all stakeholders. The limited groundwater resource is no exception. In this context, the status of the water situation of the global champion (India) of groundwater abstraction is discussed.

3.1.1 Water Availability in India – Past, Present, and Future

3.1.1.1 Prior to 1950 to Mid-1960s

In most parts of India, adequate water of generally good quality was available and the groundwater–recharge withdrawal was balanced.

3.1.1.2 Mid-1960s Green Revolution to Present

In these decades, water per capita availability declined sharply from 3000 to 1123 m^3. Currently, treated water is not accessible to ~70% of India's households. There are piped water supplies to ~40–70% of households in large cities, ~50% in small cities/towns, and 10% in rural areas for 4–5 hours a day, with 25–40% of leakage losses and illegal connections in some places.

3.1.1.3 Future Water Availability

Although considerable uncertainties are associated with the predictive models on climate, population, etc., the projected increase in global population of 5–10 billion people by 2050 is likely to decrease the per capita fresh water availability, because demand for food is expected to surge by more than 50%. If current water consumption patterns continue, going by the projected increase in population by 2025 in India, the per capita water availability is likely to drop to <1000 m^3, producing chronic "water scarcity" (Datta 2008, 2013a). As per the international norms, a region with water availability <1700 m^3 per capita per annum is "water stressed." However, superposition of climate model upon model, each with its own inaccuracy range, amplifies the overall uncertainties of current projections on how the different regions might be affected by climate.

3.1.2 Available Water Resources in India

India is endowed with abundant water resources. It has been estimated that the Ganges-Meghna-Brahmaputra Basin, covering 33% land area, has 60% of India's water resources, and the total available water is 2301 billion cubic meters (bcm) (surface water 1869 bcm + groundwater 432 bcm), utilizable water is 1123 bcm, and current use is 634 bcm. Groundwater resources have two components: static and dynamic. The static reserves (aquifer zone below the water table fluctuation zone) have been estimated as 10 812 bcm and the dynamic component as 432 bcm (Chatterjee and Purohit 2009). The available groundwater resource is about 10 times the annual rainfall (CGWB 1998). These estimates do not indicate the actual groundwater situation due to its highly heterogeneous distribution, occurrence, and quality from region to region and within parts of a region. The 2009 estimates suggested that available groundwater for irrigation was 369.6 bcm, with 71 bcm for industrial, domestic, and other purposes (Figure 3.1).

According to assessment of dynamic groundwater resources in India in March 2013, the total annual replenishable groundwater resources were 447 bcm. Keeping 36 bcm for natural discharge, the net annual groundwater availability for the entire country is 411 bcm. The annual groundwater draft is 253 bcm, out of which 228 bcm is for irrigation use and 25 bcm is for domestic and industrial use. The stage of groundwater development in the country is 62%.

It is evident that the reported estimates are different from one year to another, and there is uncertainty about the actual correct estimate. Moreover, it is not justified to take the sum of the two estimates as a measure of the total available and utilizable water resources, since significant interaction exists between shallow groundwater and adjacent surface water bodies, and the Central Water Commission (CWC) and the Central Ground Water Board (CGWB) independently estimate the separate contributions of the surface water and groundwater respectively to the overall and utilizable quantities. In most places, low annual recharge to shallow groundwater over past decades suggests limited renewal potential.

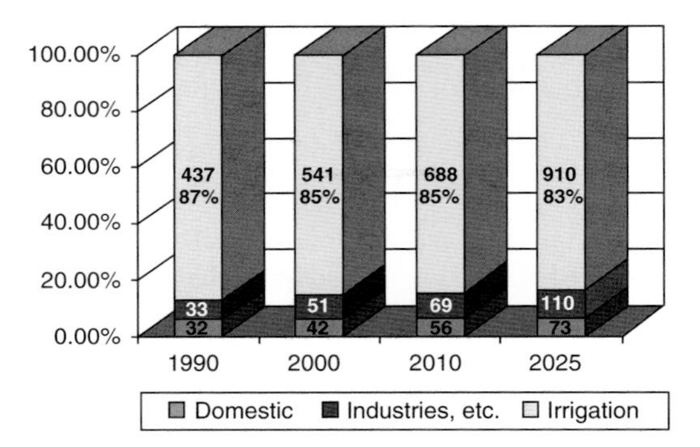

Annual water use pattern of India (bcm), and ground water availability

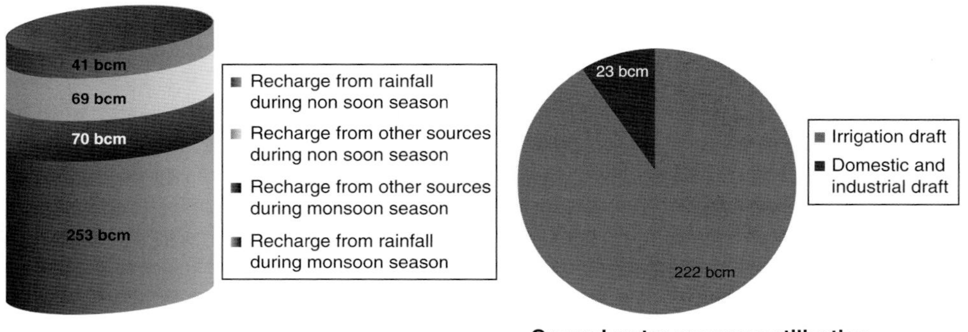

Ground water resource utilization

Figure 3.1 Annual water use pattern of India (bcm), groundwater availability, sector-wise groundwater resource draft and natural discharge in India. *Source:* CGWB (2014–2015).

3.1.3 Groundwater: Physical and Political Typologies

Groundwater is water that is present below the Earth's land surface, fully saturating the pores and cracks of the subsurface in a distinct geological formation called an aquifer, flowing slowly through the subsurface formation, depending on the formation characteristics (e.g. sand, clay, gravel, fractured rock). It is primarily formed from rainwater percolation through the soil and into the underlying geological formations, and in some places by recharge from rivers/canals/lakes and water trapped during geological processes, such as sedimentation or volcanic activity (Figure 3.2).

An aquifer is a permeable layer of underground formation, which is saturated with water and capable of yielding sufficient groundwater for human use. Aquifers can be unconfined, where the groundwater table within the aquifer layer is in direct contact with the atmosphere through soil pores, or confined, where it is over- and underlain by a semi-permeable or impermeable layer. A non-recharging aquifer, containing "fossil" groundwater, does not receive meaningful recharge on a human time scale. Political typologies of aquifers include aquifers completely contained within state boundaries of a country, national aquifers hydrologically linked with international watercourses, and transboundary aquifers shared by two or more states or countries.

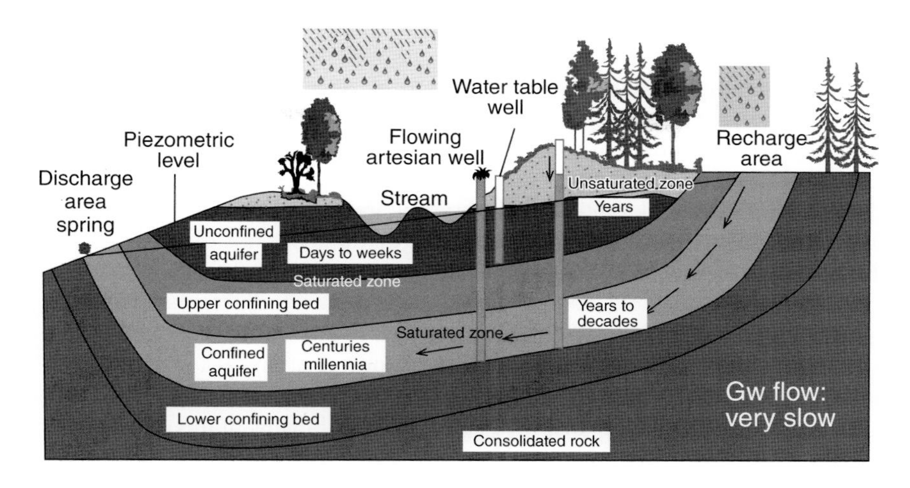

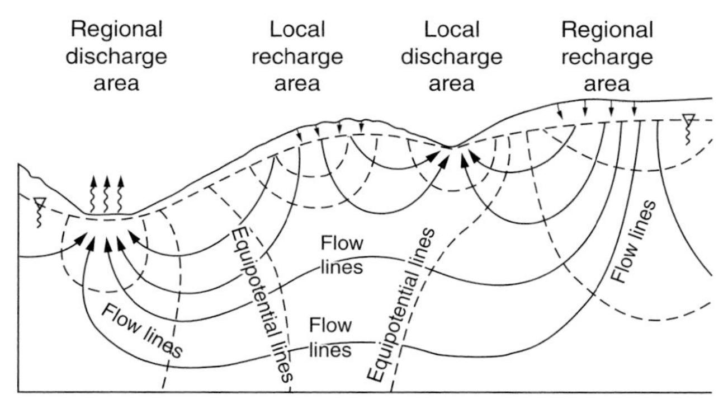

Figure 3.2 Schematic representation of general groundwater occurrence in aquifers, recharge and discharge areas, residence times and accumulation periods; flow net in a vertical section through an inclined area. (Full lines: flow lines; dashed lines: equipotential lines.)

India has a highly diversified hydrogeological set-up. Aquifer conditions vary from deep alluvial aquifers in northern areas to hard-rock aquifers in much of western, central, and southern India. The groundwater potential yield in the northern unconsolidated formations has a wide range (<10 to >40 L sec^{-1}) (CGWB 2014–2015). The groundwater potential yield in the consolidated/semi-consolidated formations has a range of 1–25 L sec^{-1}. The hard-rock aquifers are not very thick, have only dynamic groundwater resources, and groundwater depletion is common. Yet there are some areas with thick aquifers where dynamic groundwater is pumped excessively but sufficient water exists in deeper aquifers.

In the move to climate-resilient water management in the long term, managers may be confronted with important aspects such as groundwater renewal and pollution characteristics, flow velocity and direction, interaction of groundwater bodies with each other as well as with surface water, sources and dynamics of pollutants and containment of their spread from known sources, and the causes of quality deterioration under both steady and non-steady conditions.

3.1.4 India's Groundwater Crisis

India's water management is influenced and affected by decades of mismanagement, uninterested state and municipal governance, a steadily rising population (projected to be 1.7 billion by 2050), and a burgeoning middle class demanding nutrition-rich food, which all need considerable water resources. Other reasons behind the water issues in various Indian states are the lack of genuine interest at any governance level to manage water quality and quantity used; extant rules and regulations are not carried out; and poor adoption of state-of-the-art cost-effective water technologies. In spite of the fact that these problems are only likely to get worse, there are no indications that policy makers or politicians at the national level will take steps to control the rapidly increasing water crisis in the country. Politicians who come to power often seek fast but short-term solutions in the election cycles, which leads to continued water supply degeneration within the nation and increasing disputes among neighboring states over water distribution.

Water bodies in populated areas are polluted with hazardous contaminants, increasing the level of health risks for both ecosystems and humans. Not a single metropolis in the nation can offer potable water that is drinkable direct from the tap. While the quality of surface water is bad, groundwater problems are even worse, and auditing of groundwater quality is worrying. With little to no proper treatment for waste water used by industries, households and other purposes, the groundwater is steadily being polluted by contaminants. Over the past four to five decades, groundwater overexploitation has been on the rise and has become more and more unsustainable. Due to overexploitation and poor water management, groundwater levels are consistently declining across the country and in some areas, levels are dropping by more than a meter a year. In Punjab, Haryana, Rajasthan, Gujarat, Karnataka, Maharashtra, Andhra Pradesh, and Tamil Nadu, the groundwater aquifers are overstressed. Over the last 30 years, the number of private boreholes on farms has exploded due to the dearth of trustworthy irrigation, and the free electricity offered for pumping groundwater has aggravated the issue. Groundwater use increased from 58% in 2004 to 62% in 2011, and there are no signs of the rate slowing down.

Since the Green Revolution in the 1960s, groundwater exploitation in India has only risen. The lack of water has helped to advance the severe environmental, political, economic, and social issues plaguing the nation. The best way to go about creating policies for groundwater exploitation is to have dependable data and information on groundwater availability, usage, and quality. Unfortunately, groundwater is underground and effectively imperceptible. Sadly, data and information on groundwater available, usage, and quality are not consistent or dependable. India is currently using 230–250 km^3 $year^{-1}$ groundwater, which is about a quarter of world groundwater use. With groundwater being used for around 85% for domestic tasks and upwards of 60% for agriculture, India is now consuming more groundwater than the USA and China combined.

The increase in groundwater utilization can be traced back to government policies from the 1970s. At that time, external donors made free electricity for farmers a condition in return for loans to kick-start numerous agricultural development programs in order to quickly increase food production levels. While the benefits were realized in the short term, the long-term costs are the heavy financial losses that different state electricity boards are facing due to the free electricity for farmers, and the severe groundwater crisis. Official assessments released by the Indian Ministry of Water Resources reported that in 2004, 29% of groundwater aquifers were in semi-critical, critical, or overexploited condition, and that the issue was fast degenerating. More recently, in 2014, the CGWB assessed that the number of overexploited groundwater tanks had risen from 3% in 1995 to 15% in 2011. Only about 25% of aquifers are mapped so far.

The National Aeronautics and Space Administration (NASA) reported in 2009 that the Indus Basin was one of the most overstressed aquifers globally. This basin covers the states of Haryana and Punjab, India's breadbaskets. The study concluded that the groundwater reduction rate is an estimated 1 meter per three years – a 20% increment from the report made earlier by the Indian Water Ministry – and highlights the severity of India's water issues. Rising groundwater extractions come with huge water quality ramifications as well; aquifers located near the coastline are particularly exposed to sea water contamination. Additionally, there are other health risks thanks to geogenic pollution, such as arsenic poisoning. Already, these problems have arisen in some states. There will be grievous detrimental ramifications for India's environmental, energy, food, health, and water sectors if measures are not taken urgently to manage groundwater use. But if the nation continues to overexploit groundwater as it is now, by 2030 about 60% of aquifers in India will be in precarious circumstances.

3.1.4.1 Groundwater Occurrence and Availability

The groundwater situation in India is quite complex due to high heterogeneity in occurrence, diversified geological formations, climatic variations, and changing hydrochemical environments from region to region and within regions across the country. Aquifer conditions vary from deep alluvial aquifers in northern India to hard-rock aquifers in much of western, central, and southern India. In much of the north-western and south-eastern regions, deep groundwater renewal occurs via the humid climate (Datta 2013a, 2015). The public's knowledge and perception about the groundwater are limited, in terms of both distribution and quality. The groundwater resources have two components: static reserves (the aquifer zone below the water table fluctuation zone) and dynamic (Datta 2015). Significant interaction exists between shallow groundwater and surface water.

Where a perennial surface water stream intersects an aquifer, large groundwater supplies may be developed by the installation of wells or other types of subsurface intakes, which parallel the course of the stream and are at sufficient depth below the stream to permit the development of adequate gradients from the stream to the subsurface intake. As groundwater is withdrawn, water levels are drawn down in the vicinity of the intake. As pumping continues, the cone of depression deepens and the area of interception expands. When the piezometric head in the reservoir, adjacent to the underlying stream, is lowered below the stream stage, water from the stream moves down the gradient into the aquifer and towards the center of withdrawal.

The interception and diversion of surface water by induced infiltration take advantage of the slow sand filter provided by nature, as contrasted with the more widely used and more costly procedure of direct surface intake, which requires the construction of extensive filter beds. Since naturally occurring adequate filtering media do not exist everywhere adjacent to the streams, in many places the only choice is to construct filter plants. Unfortunately, knowledge of the availability and practical usefulness of groundwater reservoirs as natural filters has been rather slow to spread, and even now is not complete.

Because groundwater is invisible below the surface of the land, the public judges its availability in terms of the depth to the groundwater table and the quality in terms of color, odor, and taste. As a result, land use characteristics are determined. Moreover, policies on limiting groundwater use are not strict. Declining water tables induce change in the hydraulic gradient and can also cause seepage to groundwater from surface water sources polluted with sewage, agricultural fertilizers, and industrial by-products. Increasing groundwater use and pollution generation have exceeded the sustainable limits in many areas. Along with physical depletion, in many parts of India, groundwater is severely affected by high levels of F, NO_3, arsenic, and other harmful chemicals, exceeding the World Health Organization maximum permissible limits (MPL) for drinking water (Datta 2015). This situation is particularly problematic in rural areas with no alternative for drinking water.

Rapid urbanization and land use changes have drastically reduced the infiltration rates of rainfall into the soil and have diminished the natural recharging of aquifers. In most normal monsoon years, most of the rain falling on the land surface tends to flow as runoff, leaving very little for groundwater recharge and surface water seepage into the ground. Natural recharge measurements carried out in India suggest that shallow groundwater renewal occurred from limited modern recharge over the prior decades and from some interaction with surface water. This situation suggests that the groundwater dynamic component has limited potential for renewal. The meager natural recharge to the groundwater alone may not be sufficient to meet projected demand. In hard-rock areas, where aquifers are thin, only dynamic groundwater is available and depletion is common. There are also areas with thick aquifers where dynamic groundwater is pumped excessively, but sufficient water still exists in deeper aquifers.

Some of the important parameters of interest to planners and managers are the rechargeability of groundwater, its quantity and quality, the location of recharge intake areas, the interaction between groundwater and surface water, sources of pollution, etc. However, most groundwater development research has been fragmented and technocratic and related to groundwater flow and remediation. More research is needed on the dynamics of pollutants in the groundwater and its attenuation capacity for pollutants under natural and exploited conditions, based on a well-designed monitoring network. The pollution sources should be identified and strategies developed to contain the spread of pollution from known sources and to develop groundwater vulnerability maps. Based on these maps, potential groundwater recharge zones and protection zones need to be clearly delineated, with land use changes to restrict/eliminate unplanned waste disposal and agrochemical application in these areas (Datta 2013a, 2015).

3.1.5 Conventional and Advanced Methods of Assessing the Groundwater Situation

In view of the heterogeneity of the hydrogeological set-up normally encountered, and the complexity of the hydraulics, detailed reliable assessment of groundwater quantity and quality is very desirable through micro-level investigations for each area separately, with proper choice of technology and research on how it can be used sustainably and protected from pollution. Although data analysis by computer is a powerful tool in interpreting groundwater dynamics, whether using conventional, tracer, or satellite technologies, it is not sufficient on its own. Conventional, isotopic, and satellite methods have limitations, and available studies are too limited and fragmented to provide a reliable volumetric assessment of groundwater potential, and need confirmation under current conditions. Better assessments result from better judgments, which come from experience.

3.1.5.1 Conventional Methods of Groundwater Assessment

Generally, conventional studies on regional groundwater flow systems adopt a hydraulic approach, based on gravity-induced groundwater flow from high to low hydraulic head. However, inadequate hydraulic data on water table fluctuations, distribution, storage, etc. and the usual method of analyses provide a short-term assessment of the hydraulics. Groundwater monitoring by the CGWB started in 1969. Groundwater levels are measured four times a year during January, March/April/May, August, and November, through a network of 20 698 observation wells located all over the country. Groundwater samples are collected from these wells once a year during March/April/May to obtain information on groundwater quality changes on a regional scale. The database thus generated forms the basis for planning the groundwater development and management programs.

In India, the annual replenishable groundwater resource is 433 bcm and net annual groundwater availability is 399 bcm. The overall contribution of rainfall to the annual replenishable groundwater resource is 67%, and the share of other sources, including canal seepage, return flow from irrigation, seepage from water bodies and water conservation structures taken together, is 33%. About 73% of the annual replenishable groundwater recharge takes place during the Kharif period of cultivation. The available groundwater resource for irrigation is 361 bcm, 90% of which is utilizable quantity. The quantum of groundwater used for irrigation in the twentieth century was of the order of 128 bcm. However, estimation of groundwater recharge for a region requires a large amount of hydrological data over a long period. Therefore, these approaches are grossly inadequate and provide incorrect estimations of recharge, due to non-availability and the uncertain nature of long-term data on runoff, evaporation, transpiration, etc. Knowledge of aquifer systems and management practices that can help reduce impacts is often limited at the management level. The mechanisms that govern water demand are also not well outlined.

Volumetric estimates of groundwater recharge depend on the underground lateral and vertical extent of each of the geological formations in the assessment area. In the absence of detailed information, it is not possible to determine the relative comparison of recharge and draft of different assessment areas based on volumetric estimates. Hence, management faces various ethical dilemmas, as taking a holistic view of both the quantity and quality of the water resources, and assessment of groundwater vulnerability is of paramount importance. To optimize water use in the long term, and to protect it from depletion and pollution, some important aspects with which water resources managers are confronted include characteristics of groundwater renewal and pollution, pollutant movement and containment of spread, available fresh-water potential, etc. Conventional methods have limitations.

3.1.5.2 Isotope Methods of Groundwater Assessment

In the above-mentioned context, the potential of isotope fingerprinting of water, based on the author's own experience over four decades, has been proved to provide direct insights into the hidden aquifer system's groundwater dynamics and distribution, and has been useful for management. In India, isotope techniques have been used extensively to study groundwater provenance, occurrence, recharge and contamination characteristics, residence time (age), turnover time, flow regime, pathways of intermixing, surface water and groundwater interactions, etc.

3.1.5.2.1 Groundwater Recharge Estimation

The artificially injected tritium (^{3}H) tracer technique (Datta and Goel 1977, Datta et al. 1973, 1980a) has been extensively used to trace the downward movement of soil water over the last 40 years, and average recharge values in semi-arid and arid parts of several major river basins of India are now available. The concept of ^{3}H-tagging studies is based on the premise that downward movement of soil moisture in the unsaturated zone takes place via layer-by-layer displacement of older water by younger water above it, and this process pushes an equivalent amount of water downward, ultimately recharging the saturated groundwater zone.

Four decades of isotopic investigations tracing downward movement of soil moisture in the Indo-Ganges Plains, Sabarmati Basin, Mahi Basin, Rajasthan State, and other parts suggest that the shallow groundwater average annual recharge (Figure 3.3) from seasonal rainfall in the north alluvial plains is higher (18% in Punjab, 15% in Haryana, 20% in Uttar Pradesh) due to higher rainfall and thick unconsolidated formations conducive to recharge but limited elsewhere (<5% in highly urbanized Delhi; 1–14% in Rajasthan; 8–14% in Gujarat; 11% in the alluvial deposits of Maharashtra; 8% in Andhra

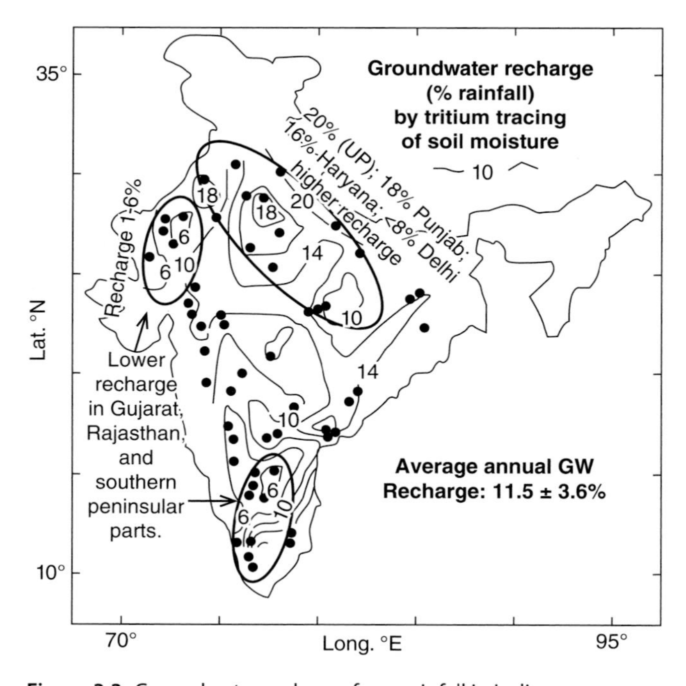

Figure 3.3 Groundwater recharge from rainfall in India.

Pradesh; <5–10% over different hard-rock areas) (Datta and Goel 1977; Datta et al. 1973, 1979, 1980a; Goel et al. 1977; Athavale et al. 1992; Datta 2013a, 2015).

Annual replenishable groundwater resource is significantly high (0.25 to >0.5 m) in the Indus-Ganges-Brahmaputra alluvial belt in north-east India covering the states of Punjab, Haryana, Uttar Pradesh, Bihar, West Bengal, and valley areas of north-eastern states, due to high rainfall and thick layers of unconsolidated alluvial formations which are conducive for recharge. The coastal alluvial belt, particularly the eastern coast, also has relatively high replenishable groundwater resources, in the range of 0.25 to >0.5 m. In western India, particularly Rajasthan and parts of northern Gujarat, which have an arid climate, the annual replenishable groundwater resources are scanty, mostly up to 0.025 m. In major parts of the southern peninsular covered with hard-rock terrains, annual groundwater recharge is <0.10 m, primarily because of comparatively low infiltration and storage capacity of rock formations in the region. The remaining part of central India has moderate recharge of 0.10–0.25 m.

More than 50% of the total geographical area of all the states put together experience medium variability, nearly 25% experience high to very high variability and nearly 20% experience low variability in rainfall. These figures coincide with medium rainfall-medium to high evaporation, low rainfall-very high evaporation, and high rainfall-medium evaporation regimes, respectively. Nearly 21% of Gujarat and 45% of Rajasthan receive less than 20 days' rainfall, nearly 51% of Gujarat and 70% of Rajasthan fall in areas which experience less than 30 days' rainfall, and nearly one-third of both states receive 30–40 days of rain a year. With regard to the other states, the area which experiences 30–40 rain days ranges from 9% to 27%; 40–50 days of rain ranges from 29% to 39%; 50–75 days of rain ranges from 27% to 58%. The Western Ghats in Maharashtra and Karnataka receive heavy rains spread over many days (>75). Both Orissa and Chattisgarh receive 50–75 days of rain in a year.

In each of the above-mentioned states, part of their geographical area falls into different rainfall variability classes (>25%, 25–30%, 30–40%, 40–50%, and >50%), and a different percentage of potential evaporation (PE) during the monsoon. In India, the major parts of the western, middle, mid-eastern and south-eastern regions are covered by water-scarce river basin systems, namely the Sabarmati, rivers of Kachchh and Saurashtra, Pennar, Cauvery, east-flowing rivers between Mahanadi and Godavari, and east-flowing rivers between Pennar and Kanyakumari. Lower rainfall coupled with higher PE increases the water scarcity. Gujarat and Rajasthan have 11% and 42% of their area, respectively, experiencing extremely low rainfall (<300 mm), and 39% and 32% respectively with low rainfall (300–600 mm). In Maharashtra, Andhra Pradesh, Madhya Pradesh, Karnataka, and Tamil Nadu, over 85% of the area experiences medium rainfall. With regard to PE, most of Gujarat and Rajasthan and 35–56% of the area of the other states have high evaporation (2500–3000 mm), with 38–65% falling into the medium evaporation regime (1500–2500 mm). Orissa and Chhattisgarh in the mid-eastern area fall entirely into the medium evaporation regime.

In view of the above hydrometeorological conditions, recharge as a percentage of direct infiltration of rainfall is low in most places, except in the near vicinity of surface water sources, indicating limited potential. However, in some places, recharge estimates are quite substantial even for the desert conditions. Although the present average percentage of groundwater recharge from rainfall might not have changed much, it varies both in space and time from region to region and within regions, depending on distribution and intensity of rainfall, evaporation, and soil properties (Datta 2013a, 2015).

In much of the north-west and south-east regions, the deep groundwater ^{14}C ages (2000–22 000 years BP) suggest limited modern renewal of groundwater by rainfall and interaction with rivers/lakes (Datta 2015); the deep groundwater renewal occurred from past humid climates. On the other hand, shallow groundwater ages (<10–20 years) suggest renewal from limited modern recharge from rainfall over the past decades, and by interaction with lakes and rivers. All these assessments suggest that the groundwater withdrawal is greater in some areas, comparable in some areas and less in some areas than previous groundwater recharge estimates by isotope methods (Datta et al. 1973; Datta and Goel 1977; Goel et al. 1977; Rangarajan and Athawale 2000; Datta 2013a, 2015). The mean natural recharge rate from rainfall over 17 major river basins is estimated to be 2.4–19.8 cm $year^{-1}$, the minimal quantity of input is about 476 km^3 $year^{-1}$, and in the Ganges Basin it is ~142 km^3 $year^{-1}$ (Rangarajan and Athawale (2000). This does not include seepage from surface water and return flow from surface water irrigation.

A comparison of recharge of different basins indicated a relatively higher efficiency of winter rains in inducing groundwater recharge (Datta et al. 1979, 1980a), and higher potential evaporation in the Sabarmati River Basin during monsoon months is likely to reduce the net groundwater recharge for a certain amount of water input. Recharge takes place insignificantly if annual water input (rainfall + irrigation) is <40 cm. Extensive (but inefficient) canal irrigation system seepage contributes ~35% of groundwater recharge. Simple mixing models of canal/river water and adjacent groundwater, developed by Datta and Tyagi (1995) and Datta and Kumar (2011) based on the Yamuna River/canal water ^{18}O and adjacent shallow groundwater ^{18}O, indicate that significant canal/river water seeps to groundwater up to 5–10 m depth of the adjacent aquifer in the Delhi area, and 2–96% river water seepage to groundwater under the floodplains at different points. However, since generally canal/river water remains polluted, it is likely to make groundwater vulnerable to contamination.

However, in arid or semi-arid zones, infiltration of excess irrigation water may contribute much more to recharge than rainfall alone. Examples are the Indus Valley in Pakistan, the Nile delta and Nile Valley in Egypt, several plains in northern China and even Mediterranean alluvial plains in Europe, especially the plain of Crau near the Rhone delta in France and many coastal plains along the Spanish Mediterranean coast and on Gran Canaria. The same is true for many oases in the foothill areas of the northern Sahara and in Central Asia, irrigated by floodwaters. In the Indus Valley (Pakistan), where a large quantity of surface water goes to irrigation, the estimated 35–40% losses of water from canals and irrigated lands account for more than 80% potential of the alluvial aquifer. In northern China, the secondary resources of many aquifers constitute a major part of the recharge: 88% in the Yinchuan plain and 51% in the Hetao plain.

3.1.5.2.2 Satellite-Based Groundwater Remote Sensing Methods

Satellite-based studies undertaken over nearly 15 years by the Indian Space Research Organization (IRSO) have assisted in building 9200 recharge structures and over three lakh wells across the country. An assessment report from the Department of Space claims that the data have been a little more than 90% accurate. A portal launched two years ago is reportedly providing continuous information that has helped in identifying prospective sites for groundwater and recharging. To conduct the mapping, the ISRO considered seven items – drainage; surface water bodies; canals; rainfall data; irrigated areas; springs; and wells – including also geological parameters such as rock type, landform, fractures/faults along with groundwater prospects with probable depth and yield range.

Based on terrestrial water storage change data from the Gravity Recovery and Climate Experiment satellites of NASA, soil-water variations data and hydrological modeling, Rodell et al. (2009) estimated that groundwater was depleted at a mean rate of $4.0 \pm 1.0\,\text{cm}\,\text{year}^{-1}$ equivalent height of water ($17.7 \pm 4.5\,\text{km}^3\,\text{year}^{-1}$) from the aquifers of Punjab, Haryana, and Rajasthan, including Delhi, between August 2002 and October 2008, and depletion was equivalent to a net loss of $109\,\text{km}^3$ of water. However, this work suggests that much of the depletion occurred within the aquifer systems of Rajasthan, where recharge is negligible. Moreover, the observation record is brief, and in the absence of any unusual trend in rainfall during 2002–2008, overabstraction of groundwater may be the cause of depletion, and, these large-scale interpretations lack the spatiotemporal resolution required to identify groundwater.

However, in contrast to documented Indian groundwater depletion due to unmanaged groundwater withdrawal, a new study using statistical analyses and simulation model results of groundwater management policy change reports regional-scale groundwater storage (GWS) replenishment through long-term (1996–2014, using more than 19000 observation locations) *in situ* and decadal (2003–2014) satellite-based groundwater storage measurements in western and southern parts of India (Bhanja et al. 2017). In parts of western and southern India, *in situ* GWS (GWS_{obs}) has been decreasing at a rate of $-5.81 \pm 0.38\,\text{km}^3\,\text{year}^{-1}$ (in 1996–2001) and $-0.92 \pm 0.12\,\text{km}^3\,\text{year}^{-1}$ (in 1996–2002), reversed to replenish at a rate of $2.04 \pm 0.20\,\text{km}^3\,\text{year}^{-1}$ (in 2002–2014) and $0.76 \pm 0.08\,\text{km}^3\,\text{year}^{-1}$ (in 2003–2014), respectively. Although such models can be useful to study impacts and trends on a regional scale, the estimated abstraction and water mass changes are uncertain, and these rarely address the priorities of stakeholders.

The water table fluctuation data from the CGWB suggest groundwater depletion in northern India of $13.2\,\text{km}^3\,\text{year}^{-1}$, which is close to the GRACE data (doi: 10.1038/nindia.2009.270). However, these analyses may not be useful for local-scale water management. New evidence from high-resolution *in situ* records of groundwater levels, abstraction, and groundwater quality reveal that sustainable groundwater supplies are constrained more by extensive pollution than depletion (MacDonald et al. 2016). Groundwater volume up to 200 m depth has been estimated to be >20 times the combined annual flow of the Indus, Brahmaputra, and Ganges, and the water table was stable or rising across 70% of the aquifers between 2000 and 2012. Groundwater levels are falling in the remaining 30%, amounting to a net annual depletion of $8.0 \pm 3.0\,\text{km}^3$. Within 60% of the aquifers, groundwater is polluted by excessive salinity/arsenic. More confident analysis of the estimates of groundwater storage changes requires additional data and improved groundwater recharge and level monitoring networks, instead of relying on uncertain, remotely sensed data.

3.1.5.3 Isotope-Based Assessment of Groundwater Provenance and Occurrence

Groundwater availability differs from location to location, characterized by specificity of occurrence. Evidently, indiscriminate groundwater use has crossed sustainable limits, and due to variations in social, economic, and political factors, no single template for management can be developed. Hence, it is desirable to better understand the occurrence and characteristics not only of shallow and fresh groundwater resources but also for groundwater at great depths and for saline and brackish groundwater, as a potential water resource. Instead of relying on previously collected data from multiple sources, the new dataset should include groundwater chemistry, hydrogeological characteristics, and horizontal and vertical extents of principal aquifers (regionally extensive aquifers or aquifer systems which have the potential to be used as a source of potable water), and groundwater use.

Groundwater provenance and occurrence were investigated via isotope techniques, taking advantage of the "memory" maintained by the rainfall-stable isotopes ^{18}O and ^{2}H under recharge conditions on reaching the groundwater. The groundwater-stable isotope composition is a proxy indicator of the isotopic composition of the rainfall by which groundwater has been recharged. The ^{2}H and ^{18}O isotopes are conservative tracers (i.e. influenced solely by convective transport) and remain constant in the groundwater flow direction, unless mixed with water of different isotopic composition and subjected to evaporation during recharge.

For example, in the Delhi area, depleted ^{18}O is generally associated with heavy rainfall, and rainfall-deficient years are associated with relatively enriched ^{18}O in monthly rainfall; variation in recharge from location to location, both in space and time, results in a wide range of ^{18}O isotopes in groundwater, both laterally and vertically, suggesting an inhomogeneous system, and groundwater renewal has a selection effect in favor of isotopically depleted heavy rainfall (Datta et al. 1991, 1996a). The areas with small isotope gradients helped to delineate the potential groundwater recharge zones and hydrodynamic zones, which were useful in locating sites of wells with high yield and efficiency. The groundwater ^{18}O and Cl distributions suggested two main flow systems vertically in Delhi (Datta et al. 1994; Datta and Tyagi 1995): (i) uppermost local flow, rapidly circulating, low salinity, more vulnerable to overexploitation, and (ii) relatively slow circulating intermediate zone, more vulnerable to salinization and depletion.

3.1.5.4 Stratification of Groundwater

The ^{18}O isotopic data of groundwater showed that stratification exists in the Delhi area (Datta et al. 1994, 1996a, b, 1999). Radiocarbon ages and ^{18}O data indicated significant stratification in the groundwater of the Pushkar Valley (Ajmer), Rajasthan, Sabarmati Basin, Gujarat (Borole et al. 1979; Datta et al. 1994, 1980a, b) and Jaisalmer District, Rajasthan (Datta et al. 1980b). Straight-line relationships between groundwater ^{18}O and Cl in the Delhi area indicated that groundwater intermixing takes place along specific flow pathways (Datta et al. 1994, 1996a). Using simple mixing equations, the lateral component of recharge is estimated to range from 25% to 70%, influenced by the flow pathways of intermixing and the extent of the hydrodynamic zones (as indicated by small isotopic gradients).

Another example in Rajasthan (area 342 239 km^2) has 60% of the area under arid conditions with average rainfall <400 mm while 40% is semi-arid with 550–800 mm average rainfall. Deficient and extremely erratic rainfall is insufficient not only for rain-fed agriculture but also for developing surface water resources, and very adversely affects the recharge of aquifers, wells, and other water bodies, causing severe water scarcity. Although the rainwater mostly tends to run off down the dune slopes, shallow aquifers in the alluvial plains, interdune plains, piedmont plains, and intermontane basins have the capacity to retain sufficient groundwater, and allow low-level agriculture in areas where the sand dunes have stabilized and support mature soils. Deep aquifers are generally present in the hard-rock formations of piedmont plains, and the rocky/gravelly piedmont or limestone and sandstone buried piedmonts covered by colluvial alluvial sediments. Groundwater occurs at 30–120 meters below ground level (mbgl), and the water table has declined by 10–40 cm $year^{-1}$.

The underground flow of the Ghaggar-Saraswati River is essentially groundwater flowing along the sands and sediment deposits of the river channels or even through subsurface bedrocks, replenished by monsoon water from the channels and surrounding plains. Very deep-seated aquifers are generally found in sandstone formations under thick cover of dune sand and other formations in Jaisalmer, Bikaner, and northern Barmer and Churu districts, and have either fossil water or their sources of recharge far away from the aquifers. A number of environmental threats, such as a large variation in rainfall, inadequate availability of good-quality water, droughts, desertification, etc., have caused unexpected problems for groundwater sustainability.

Environmental radioactive isotope ^{3}H (half-life ($T_{1/2}$) = 12.3 year; produced in the upper atmosphere by cosmic radiation), being short-lived, can help in assessing groundwater provenance. Absence of ^{3}H in groundwater indicates that the water is not of recent origin. ^{3}H concentration is expressed in tritium units (1TU = 1 ^{3}H in 10^{18} ^{1}H atoms). In the Jalore area of Rajasthan, shallow groundwater, near the river course and away from it, had ^{3}H values of 5–20 TU and 1.4–3 TU respectively, suggesting modern recharge (Sinha and Navada 2008). Deep groundwaters >50 m away from river course had depleted ^{2}H and ^{18}O and negligible ^{3}H, indicating absence of modern recharge.

3.1.5.5 Groundwater Residence Time/Age

Groundwater residence time/age is estimated using ^{14}C ($T_{1/2}$ = 5730 year), produced naturally in the upper atmosphere by cosmic radiation. Since ions of carbonates and bicarbonates dissolved in water have radioactive ^{14}C, carbon in water may exchange with carbon in the aquifer matrix, and ^{14}C ages are determined accordingly. The dug well waters in Jaisalmer have a ^{14}C age of 5000–1800 years BP, and the tube well waters have a ^{14}C age of 22 000–6000 years BP, with negligible tritium, indicating absence of modern recharge, and the groundwater is paleo-water recharged during pluvial phases in the Holocene (Rao 2003). The ^{14}C age was ~20 000 years for flowing well water in southern Haryana and ~8000 years for tube well waters (depth ~20 m) (Kulkarni et al. 1989). A ^{14}C age of 22 000–2000 years BP suggests that the residence time of water in the aquifer is long after it was recharged from rainfall. Such "fossil" water reserves can possibly help to minimize depletion by unsustainable exploitation.

3.1.6 Isotope-Based Assessment of Groundwater Vulnerability to Contamination

Groundwater quality is declining everywhere and an alarming picture has emerged in many parts of India. In many of the investigated areas, public ignorance about environmental considerations, lack of provisional basic social services, indiscriminate disposal of increasing industrial and domestic wastes on land, into rivers and unlined drains, mining activity, and unplanned application of agro-chemicals and improperly treated sewage water have resulted in excessive accumulation of pollutants on the land surface. Subsurface leaching of contaminants from landfills as well as seepage from canals/rivers and drains has caused severe degradation of groundwater (Datta and Tyagi 1995; Datta et al. 1996a). Most groundwater quality problems are caused by contamination, overexploitation, or a combination of the two, which are all difficult to detect and resolve.

Decontamination solutions are usually very expensive and time consuming and not always effective. The natural chemical content of groundwater is influenced by depth of soils and subsurface geological formations with which groundwater remains in contact. Shallow groundwater is mainly of calcium bicarbonate and mixed type, and is generally suitable for different types of use. However, other types of water, including sodium chloride water, are also available. The water quality in deeper aquifers also varies from place to place and is generally suitable for common purposes. Salinity problems exist in coastal areas.

In large parts of Delhi, Punjab, Haryana, Gujarat, Maharashtra, Karnataka, Orissa, Andhra Pradesh, Assam, Bihar, Chhattisgarh, Jharkhand, Jammu and Kashmir, Kerala, Uttarakhand, Madhya Pradesh, Rajasthan, Telangana, Tamil Nadu, Uttar Pradesh, and West Bengal, groundwater is severely vulnerable to salinization and pollution with F ($<1-45\,mg\,L^{-1}$) and nitrate ($<20-1800\,mg\,L^{-1}$) heavy metals, arsenic, etc. exceeding the WHO maximum permissible limit (MPL) in drinking water (Datta and Tyagi 1995; Datta et al. 1996a, b, 1997, 1999; Datta 2015). Elevated levels ($>0.05\,mg\,L^{-1}$) of arsenic in groundwater were reported from parts of West Bengal, Assam, Bihar, Jharkhand, Uttar Pradesh, Punjab, Haryana, Chhattisgarh, Karnataka, and Manipur (www.searo.who.int/india/topics/arsenic/en).

Trace to excessive amounts of heavy metals, such as Zn ($3-41\,\mu g\,L^{-1}$), Cu ($5-182\,\mu g\,L^{-1}$), Fe ($279-1067\,\mu g\,L^{-1}$), Mn ($<1-76\,\mu g\,L^{-1}$), Pb ($31-622\,\mu g\,L^{-1}$), Ni ($<1-105\,\mu g\,L^{-1}$), Cd ($<1-202\,\mu g\,L^{-1}$), have been found in the groundwater in some parts of Delhi, Haryana and Uttar Pradesh near industrial sites, Andhra Pradesh, and Madhya Pradesh (Datta et al. 1999). In coastal areas of Gujarat, sea water intrusion has affected groundwater quality (Rina et al. 2011, 2013). Highly skewed ranges of groundwater Cl, NO_3, F, and heavy metals associated with enriched ^{18}O indicate pollution by infiltration of evaporated surface runoff and rainwater along with anthropogenic pollutants in soil and wastes from both point and non-point sources (Datta et al. 1996a, b, 1999; Rina et al. 2011, 2013). In some areas of Delhi, the absence of known major geological sources of fluoride and nitrate in the region, excessive application of fertilizers and discharges from steel, aluminum, brick and tile industries, barnyard and silo wastes, and disposal of crop residues are major causes of pollution (Datta et al. 1996b). In the adjacent areas of the River Yamuna in Delhi, total and fecal coliform bacteria counts in groundwater vary from 1 to 570 000 and 1–420 MPN/100 mL respectively (Datta and Kumar 2011).

Variable mixing of contemporary recharge with old groundwater, degrees of recharge/evaporation, adsorption/dispersion processes in the soil zone, slow infiltration of agricultural and urban surface runoff, and overexploitation-induced lateral intermixing of contaminated groundwater with fresh water along specific flow pathways result in variation in pollutant contents, increasing lateral extension of contaminated groundwater and decreasing the fresh-water potential (Datta et al. 1996a, b). The spatial distribution and relationships of isotope ^{18}O and chemical contaminant species clearly indicate the direction of groundwater movement and mixing of multiple sources of highly saline/

contaminated groundwater with relatively fresh groundwater or river water along specific flow pathways (Datta and Tyagi 1995; Datta et al. 1996a). The conceptual model (Datta et al. 1979, 1980a) is useful in predicting groundwater pollution, caused by seepage of irrigation waters (containing agrichemicals, etc.), and in generating safe water (Soni et al. 2009a, b).

In most parts of India, moderately to highly saline groundwater occurs at all levels, and EC: 200–3000 μmhos cm^{-1} is observed in shallow groundwater. Brackish and saline to highly saline groundwater (EC: 3000–16 000 μmhos cm^{-1}) exists at all levels in north, north-western and south-western areas, and groundwater salinity increases with depth. Brackish groundwater as an unconventional water source may offer a partial solution to current and future water demands. Previously published data relating to brackish groundwater resources are limited to a few regional- and state-level studies. Hydrochemical data includes concentrations of total dissolved solids (TDS), major ions, trace elements, nutrients, radionuclides and physical properties of the water (pH, temperature, and specific conductance). The database should provide data on well location, yield, depth, and contributing aquifer.

Groundwater vulnerability to contamination is related to its age. Groundwater that was recharged prior to the Holocene is widely assumed to be unaffected by modern contamination. However, analysis of isotope (^{12}C, ^{13}C, ^{14}C, ^{3}H) data of "fossil" groundwater from 6455 wells around the globe suggests that fossil waters make up 42–85% of total aquifer storage in the upper 1 km of the Earth's crust; most waters are pumped from wells >250 m depth, and 50% of wells, which are dominated by fossil groundwater, also contain detectable levels of tritium (Jasechko et al. 2017). This suggests presence of decadal-age waters, and contemporary contaminants may be able to reach deep fossil aquifers. Since fossil groundwater would take millennia to replenish once used up, considering water quality risk is important along with sustainable use in the management of fossil groundwater resources.

India is the largest user of groundwater in the world, and groundwater will continue to be used intensively. Dependency on groundwater becomes particularly acute during summer, when demand for water for different purposes increases dramatically in relation to an inelastic supply. For some parts of India, large-scale development of groundwater has caused decreases in the amount of water present in aquifer storage and discharges to surface water bodies. Water supply in some areas, particularly in arid and semi-arid regions, is not adequate to meet demand, and severe drought is affecting large parts of India. Future water demand is projected to heighten the current stress on groundwater. This combination of factors led to concerns about fresh-water availability. Since hydrogeological, social, economic, cultural and political factors, aggregated impact of millions of individual pumping decisions, and emerging groundwater problems can vary greatly at local or regional scales, no single template for management can be developed. Large-scale development of publicly funded tube wells tends to be supply driven; it is difficult to police adequately legal regulatory provisions at the national level.

3.1.7 Causes of Increasing Water Demand

Water resources have a highly heterogeneous distribution from region to region and within regions. Increasing demand is mainly due to inadequate availability and zonal disparity in surface water supply, rapid growth in population, agricultural intensification, urbanization, industrialization, and competition for economic aspirations, leading to conflicts among different sectors. Out of the annual water withdrawal, agriculture uses ~80–90% followed by domestic water demand (~7%) and industrial sectors (~2%). The urban clusters demand moderate quantities of high-quality water from the unreliable infrastructure created by state water departments and municipal supply utilities. In rural areas, where 70% of the population lives, people demand large quantities of high-quality water from inefficient distribution systems. The meager natural recharge to groundwater alone may not be sufficient to meet demand.

3.1.8 Status of Water Scarcity

Traditionally, cities facing increased demands for water, along with variable supply, have relied on large-scale, supply-side infrastructural projects such as dams and reservoirs. This is essentially "supply-side" management but this type of management is costly in political, economic, and environmental terms. Economically, water has to be transported over long distances, increasing the costs of supply. Additionally, the water is often of inferior quality and so requires additional treatment for potable consumption, increasing energy as well as chemical costs in water treatment plants. Environmentally, large-scale diversion of water disrupts the health of waterways and aquatic ecosystems.

Despite abundant water resources, many cost-intensive water development projects, river action plans, etc., about half of India's population lives in areas of physical water scarcity (i.e. demand > supply). The rest face economic water scarcity due to institutional, financial, human factors, etc. at different times, due to variation in the actual availability and distribution of water over space and time (Datta 2014). While some areas receive 650 liters per capita per day (lpcd), some areas only 25 lpcd. In urban areas, 135 lpcd demand is three times the rural demand of 40 lpcd. However, the per capita water availability index may not be valid with regard to socioeconomic disparities in water usage.

3.1.8.1 Common Reasons for Water Scarcity

The situation of each region/city/village may vary, but the common reasons for water scarcity at a watershed scale are erratic spatial and temporal distribution of annual rainfall (average: 50–100 cm), increasing demand, zonal disparity in water supply, ineffective implementation of provisions for efficient supply, unregulated groundwater extraction and use, unplanned wastes disposal causing moderate to high pollution of surface water and groundwater, lack of institutional reforms, interstate river disputes, periodic droughts, inadequate knowledge, and ethical framework (Datta 2005, 2015). Another reason is that only 31% of urban waste water is treated per day; out of 815 sewage treatment plants (STPs), 522 are operational, 145 are under construction and 79 do not work; 75% of water pollution is caused by untreated domestic sewage discharge. Water scarcity is also due to moderately to highly polluted 70–80% surface water and groundwater resources, which are unsafe for drinking/farming in many areas.

In recent decades, water resource managers around the world have embraced real-time reporting of continuous records of water levels and discharge. Highly responsive and adaptive decisions with many improved outcomes for water supply management are based on what is happening, and not on what has happened or what could happen.

Laboratory analysis of a water quality sample links a lot of data to a singular point in time and space. However, the objectives for monitoring may span spatial and temporal scales from point sampling to watershed assessment (e.g. to characterize waters, identify trends, assess threats, inform pollution control, guide environmental emergency responses, and support the development, implementation, and assessment of policies and regulations). Flow and water quality are highly correlated as a result of integration of sources of varying water constituents, dilution and concentration. The integration of all contributing waters into flow at a specified location inherently connects point-scale observations to watershed-scale events.

To reduce demand for scarce water, cities are turning to demand management strategies to make better use of existing supplies before plans are made to further increase supply, where demand management is the promoting of water conservation during times of both normal conditions and uncertainty, through changes in the practices, cultures and people's attitudes towards water resources. In addition to the numerous environmental benefits of preserving the health of ecosystems and their habitats, demand management is cost effective as it allows cities to better allocate scarce financial resources, which would otherwise be required to build expensive dams and water transfer infrastructure.

3.1.9 Reasons for Overdependence on Groundwater

Large-scale, publicly funded tube well developments tend to be supply driven; legal and regulatory provisions at the national level cannot be policed adequately. The sustainability of long-term water use is compatible with the limited depletion of aquifer reserves only in the short term. Therefore, management choices are between two generally acceptable approaches: (i) optimal yield; that is, merely tapping an often minimal renewable resource, allowing for deliberate short-term limited use of ground-water between recharge events, and (ii) controlled exploitation of a more plentiful non-renewable resource until it is exhausted, recognizing the necessity to promote socioeconomic development in areas receiving very limited natural recharge but introducing water conservation measures (Datta 2005).

Although efforts are being made for the harvesting of rainwater and groundwater recharge, in highly urbanized areas with very high population density, such as Delhi, rooftop harvesting cannot provide all the required water at all times of year. Moreover, groundwater and surface runoff in such areas are highly polluted. If such structures are not maintained properly, water quality may be a problem. Moreover, indirect groundwater recharge enhancement may work for shallow level circulation, but deeper systems require sophisticated injection and alternative sources of high-quality water. When there is intensive development of groundwater, a significant proportion of the interannual renewable resource is withdrawn from the aquifers, which in turn noticeably modifies their hydrogeological functioning, and causes significant ecological, political or socioeconomic impacts, or important changes are produced in river and adjacent aquifer relationships.

The intensive use of groundwater, mostly but not exclusively developed in recent decades in arid and semi-arid countries, has produced a large number of benefits for society. These include the affordable supply of drinking water and the development of irrigated land, which have contributed to health improvement and famine alleviation for hundreds of millions people in developing countries. The large water storage capacity of aquifers allows users to manage interannual precipitation variability. Aquifers become an efficient solution to mitigate drought impacts. The guarantee in supply, coupled with the low cost of extraction of groundwater facilitated by scientific and technological advances, have led to a spectacular increase in groundwater use, especially for irrigation, in many arid, semi-arid, and coastal areas.

Groundwater development significantly increased during the second half of the last century in most semi-arid or arid countries. This was mainly undertaken by a large number of small (private or public) developers, and often scientific or technological control of this development by the responsible water administration was inadequate. In contrast, the surface water projects developed during the same period are usually of larger dimension and were designed, financed, and constructed by government agencies which normally manage or control the operation of such irrigation or urban public water supply systems. A usual axiom is that groundwater is a fragile resource that should only be developed if it is not possible to implement conventional large surface water projects.

Increasing demand and socially intrusive, ecologically damaging, capital-intensive water resources development projects have forced people to depend more and more on groundwater. Globally, India is the largest user of groundwater, and it will continue to be used intensively. Dependence on groundwater increased due to its apparent reliability, easy accessibility, availability and flexibility of exploitation by tube wells, technological developments in construction of deep tube wells, water abstraction devices and methods, provision of subsidized electricity for pumping in many states, and easy credit availability from financial institutions. However, from the information provided by water authority reports, it is difficult to assess exactly the groundwater response to land cover changes, agricultural activities, industrialization, population growth, and climate.

Due to existing socioeconomic factors, many of the irrigation developments in which groundwater is involved have been carried out by the final users, with almost no planning or control by public or governmental agencies, which often do not have the necessary human and economic means, or the know-how, to cope with this new situation. Now, groundwater supports over 80% of rural and urban water supplies, and ~60–90% of irrigated agriculture through wells owned and controlled by farmers, yet they remain indifferent to its quality and renewability, unless the water is saline and water table has declined considerably. To a significant extent, groundwater overabstraction, depletion, and degradation are due to lack of well-outlined actions to govern water demand and insufficient policy frameworks. Defining the groundwater footprint (the area required to sustain groundwater use and groundwater-dependent ecosystem services) shows that humans are overexploiting groundwater in many large aquifers that are critical to agriculture.

3.1.10 What is Overexploitation of Groundwater?

The term "overexploitation" has been frequently used during recent decades. Nevertheless, the concept of aquifer overexploitation is poorly defined and resists a useful and practical definition. A number of terms related to overexploitation can be found in the water resources literature; examples include safe yield, sustained yield, perennial yield, overdraft, fossil groundwater exploitation, groundwater mining, optimal yield, etc. In general, these terms have in common the idea of avoiding "undesirable effects" from groundwater development. However, "undesirability" depends mainly on the social perception of the issue, and is more related to the legal, cultural, and economic background of the region than to hydrogeological facts.

The historical situation has often produced two effects: (i) most water administrations have limited understanding and poor data on the groundwater situation and value; and (ii) in some cases the lack of control on groundwater development has caused problems such as depletion of the water level in wells, decrease of well yields, degradation of water quality, land subsidence or collapse, interference with streams or surface water bodies, ecological impact on wetlands, etc. These problems have been magnified or exaggerated by groups with lack of hydrogeological know-how, professional bias or vested interests. Because of this, in recent decades groundwater overexploitation has become a kind of "myth" and groundwater mining, that is, the development of non-renewable groundwater resources, is also considered "overexploitation."

In general, overexploitation is only diagnosed *a posteriori*. To analyze *a priori* the likelihood of an aquifer becoming stressed (or overexploited), five main effects or indicators are considered:

- decline in water levels
- deterioration of water quality
- land subsidence
- hydrological interference with streams and lakes
- ecological impact on aquatic ecosystems fed by groundwater.

Observation of a trend of continuous significant decline of the levels in water wells over one or two decades is frequently considered as a clear indication of imbalance between abstraction and recharge.

3.1.11 Reasons for Unbalanced Groundwater Recharge and Withdrawal

Water withdrawn artificially from an aquifer is derived from a decrease in storage in the aquifer, a reduction of the previous discharge from the aquifer, an increase in the recharge, or a combination of these changes. The decrease in the discharge plus the increase in recharge is termed "capture," which may occur in the form of decreases in the groundwater discharge into streams, lakes, and the ocean, or from decreases in that component of evapotranspiration derived from the saturated zone. After a new artificial withdrawal from the aquifer has begun, the head of the aquifer will continue to decline until the new withdrawal is balanced by capture. In many circumstances. the dynamics of the groundwater system are such that long periods of time are necessary before any kind of equilibrium conditions can develop. Some examples show that the time necessary to reach a new equilibrium or steady state between groundwater extraction and capture may take decades or centuries.

3.1.11.1 Groundwater Depletion in India

When groundwater is pumped from the water table aquifer at rates sufficient to lower water levels below the piezometric surface of the artesian formation, leakage gradients will be reversed and groundwater now percolates upward from the artesian formation and enters the watertable aquifer if the confining bed is permeable enough. In areas of extensive development, where a large regional lowering of groundwater level may occur, vertical leakage may be induced from underlying or overlying formations at appreciable rates and over a quite large area.

Under these conditions as development continues and increases in magnitude, the contribution of groundwater from adjacent formations also increases. If the groundwater contained in overlying or underlying formations is of different chemical composition from that in the aquifer in which development occurs, there will result noticeable changes in the quality of groundwater pumped from the aquifer. The rate of vertical leakage is proportional to the rate of pumping from wells in the aquifer. Thus, it would appear that changes in chemical quality would be most marked in wells or areas where withdrawals are greatest. Although this generalization may be appropriate for many wells, for some wells it may be modified by other conditions such as differences in the degree of penetration of the pumped wells, differences in the vertical permeability from area to area, and the effects of differences in density of the several groundwaters involved.

Groundwater being a "common property" resource under state government jurisdiction, competition remains among users to extract as much water as each one can. In rural areas, farmers depend on groundwater due to the apparent reliability and flexibility of groundwater exploitation by the wells owned and controlled by the farmers, and they overexploit groundwater for irrigation. In the political arena, in some parts of India, electricity subsidies provided by the government to farmers to pump groundwater are driving unfettered groundwater mining even during good monsoon years. However, farmers remain indifferent about the quality unless the groundwater is saline. While electricity use for pumping groundwater facilitates abundant grain production, its own production requires water in hydroelectric power plants. This low-cost irrigation encourages users to draw more water, resulting in lowering of groundwater and reduction in the productivity of wells.

To meet the requirement for water, millions of private tube wells have been constructed. The groundwater level in many basins is declining due to unregulated overexploitation even when recharge conditions are good, placing a burden on farmers in terms of increasing drilling and extraction costs. The situation is already critical in northern Gujarat, southern Rajasthan, coastal Tamil Nadu, parts of Haryana and Punjab. But much more insidious is the impact on water quantity, due to the fact that in India, the area under irrigation has grown from <20% in the 1950s to over 50% at present. Water tables are dropping at alarming rates in nearly all parts of India and water availability per person has dropped precipitously as well. The selection of water-intensive cash crops (e.g. sugarcane and cotton) is an additional factor.

In many agriculturally productive regions of India, where a large number of small farmers (with land holding <2 ha) exist, groundwater is pumped either by electricity or by dieselized pumps to serve agricultural needs. The increasing price of diesel will also affect the food price index. For example, in Uttar Pradesh, which is part of the Indo-Gangetic plains with extensive agriculture, where the majority of farmers have >2 ha of land holding, the number of diesel pump sets in 1979–1980 was 581 998, rising to 2 965 357 in 1998–1999, a fourfold increase. With this trend, in the intervening years the number of diesel pump sets would have increased significantly. The increased use of diesel to run these sets would definitely be costing more under the depleting water-level conditions. According to estimates from the Ministry of Water Resources, under the

Fresh Water Year 2003 project, in alluvium areas, around 0.40 kW h^{-1} more energy will be required for a 1 m decline in groundwater level. Hence, if this is the case in only one state of India, one can imagine the rise in cost of food items or energy requirements for proper irrigation of food crops.

The above analyses suggest that for enhancement of food production, besides high-yielding genetic varieties and usual agronomic practices, proper management of the limited available groundwater resource is important. In India, since rural population density is high and most agricultural holdings are small and highly fragmented, social equity must be ensured in terms of water availability. Water management practices may be based on increasing water supply and managing water demand under the scarce water availability conditions. Much of the groundwater overexploitation is taking place in arid/semi-arid northern, western, central and southern India, areas that receive low/medium rainfall and have a low recharge. This dependence on groundwater is causing rapid depletion and degradation in the resource base. In India, around 29–70% of groundwater blocks are semi-critical, critical or overexploited (CGWB 2014–2015; Planning Commission 2007a, b).

In western India, particularly the arid regions of Rajasthan, annual recharge is scanty and groundwater draft is low (up to 0.04 m) but in the arid parts of Punjab and Haryana, draft is relatively high (0.16–2.5 m). The present groundwater development is ~106% in Delhi, 94% in Punjab, 84% in Haryana, 60% in Uttar Pradesh, 41–51% in the western states, 17–30% in the central states, and 24–60% in the southern states. Overabstraction runs at 100–260% in some parts of Punjab, Haryana, Gujarat, and Rajasthan. During recent decades, due to annual recharge and withdrawal imbalance, the water table has declined by 2–8 to 30–40 m in different parts of the Delhi area, by 50–60 m in Punjab, by 1–10 m in north-west Uttar Pradesh, 3–8 m in Haryana, 7–10 m in Rajasthan (up to 20–60 m in the Thar Desert), 40–60 m in Gujarat, and >4 m in other states (CGWB 1998; Datta 2005, 2013a, 2015, 2017).

According to the 2014–2015 Groundwater Year Book (CGWB 2014–2015), in India, the water level fluctuation of pre-monsoon 2013 to pre-monsoon 2014 shows that out of 11 851 wells analyzed, 7638 (64%) had risen and 3764 (32%) had a fall in water level. The remaining 449 (4%) stations did not show any change in water level. About 47% of wells had <2 m rise in water level. About 10% had a 2–4 m rise in water level and 7% had >4 m rise. About 32% of wells had a decline in water level, out of which 25% had <2 m decline. About 4% of wells had a 2–4 m decline in water level. Only 3% of wells had >4 m decline in water level range. The majority of the wells had a 0–2 m rise/decline in water level. The water level fluctuation from January 2014 to January 2015 (Figure 3.4) shows that out of 12 822 wells analyzed, 4592 (36%) wells had risen and 7960 (62%) had a fall in water level. The remaining 270 (2%) wells did not show any change in water level. About 29% of wells had <2 m rise, about 4% had a 2–4 m rise, and 3% had >4 m rise. About 62% of wells had a decline in water level, out of which 47% had <2 m fall. About 10% of wells had a 2–4 m fall in water level. Only 5% had >4 m fall.

A comparison of depth to water level between August 2014 and August 2004–2013 decadal mean indicated that about 49% of wells had risen in water level, out of which 39% had <2 m rise. About 8% of wells had a 2–4 m rise in water level and about 3% had >4 m rise. About 50% of wells had a decline in water level, out of which 35% had a 0–2 m decline. Nine percent of wells showed a decline in water level in the 2–4 m range and the remaining 6% were in the range of >4 m. Decline in water level of >4 m was prominent

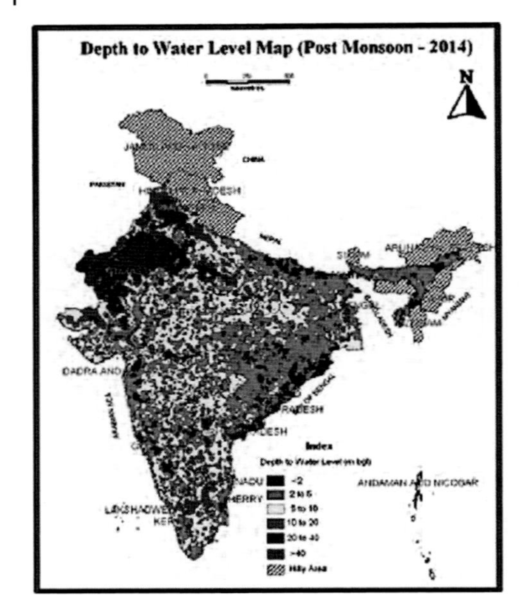

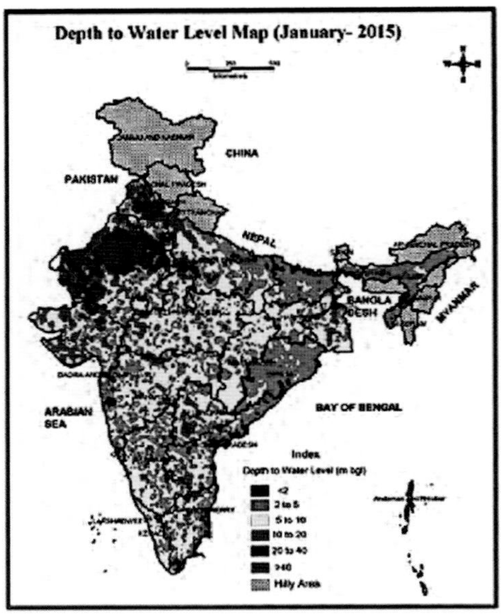

Figure 3.4 Depth to groundwater level in India. *Source:* CGWB (2014–2015).

in the states of Punjab, Haryana, Delhi, Rajasthan, Chandigarh, Gujarat, Tamil Nadu, Telangana, Andhra Pradesh, Karnataka, Maharashtra, and Madhya Pradesh. Rise in water level of >4 m was observed mostly in the states of Chandigarh, Gujarat, Himachal Pradesh, and Rajasthan. The remaining 53 (<1%) stations did not show any change in water level. Maximum fall was observed in and around parts of Punjab, Rajasthan, and Tamil Nadu. A rise in water level was observed in almost all parts of the country, and as patches in Punjab, Kerala, and Tamil Nadu.

According to the 2014–2015 Groundwater Year Book (CGWB 2014–2015), India has an estimated 30 million groundwater structures. Over 90% of groundwater is extracted for irrigation for 60% of the irrigated area, underscoring India's dependence on groundwater irrigation. Over the last four decades, around 84% of the total addition to the net irrigated area has come from groundwater, according to the CGWB. Across India, barely 3% of wells registered a rise in water level exceeding 4 m in the year ending January, according to this report. Only 35% of wells showed any rise in water level, which declined in 64% of wells. Average water levels in January were lower than the average water level between 2006 and 2015.

An inconsistent pattern of rise or fall in water level in a region or parts of a region suggests that the regional-scale aquifer systems are being affected by ineffective governance. Due to the absence of any pricing mechanism and strict regulation, and with large-scale tube well installations being supply driven, indiscriminate groundwater exploitation and its wasteful utilization have continued, out of balance with natural recharge and crossing sustainable limits (Datta 2005, 2013a, 2015, 2017). It is evident that both the groundwater draft as well as the stage of groundwater development is more than 100% in the states of Haryana, Punjab, and Rajasthan, while recharge is low in arid areas (Chatterjee and Purohit 2009). These estimates have been generated based on volumetric estimates of annual groundwater recharge and draft divided by the area of the assessment unit to arrive at estimates per unit area.

According to a 2012 UNESCO report, the trend of falling water levels in India is 251 km^3 annual groundwater extraction rate, which is equivalent to 26 times the water stored in the Bhakra Dam, making India the world's biggest consumer of groundwater. The problem and the advantage with groundwater is its decentralized access. A license is only needed to sink a well on owned land and extract water. Consequently, in Punjab, Rajasthan, and Haryana, groundwater abstraction exceeds the rate at which it is being replenished through rainfall, backflows from irrigation and seepage from canals, other water bodies, and conservation structures. A license does not prevent groundwater exploitation, and also breeds corruption within the system. Thirty million groundwater structures cannot be policed through a license quota.

3.1.12 How to Determine Groundwater Safe Yield

The concept of "safe yield" is used to determine how much water can safely be withdrawn from an aquifer system, so that annual withdrawals do not exceed the annual rate of recharge. However, as human activities change the system, the components of the water budget (inflows, outflows, and changes in storage) also will change and must be accounted for in any management decision. "Basin yield" can be defined as the "maximum rate of withdrawal that can be sustained by the complete hydrogeological system in a basin without causing unacceptable declines in hydraulic head anywhere in the system or causing unacceptable changes to any other components of the hydrological cycle in the basin." The following simple relationship describes a natural aquifer system, not influenced by human activities:

$$\text{Steady state of dynamic equilibrium: inflow} = \text{outflow} \pm \Delta\ \text{storage.}$$

In the undeveloped aquifer system, the inflow component is mainly represented by precipitation/recharge. The outflow component is composed of discharge into boundary waters such as the ocean, lakes, and other boundary features. It may also include discharge into surface waters such as streams and rivers that run through an aquifer watershed. Recharge of the aquifer may also increase, often in the form of "induced recharge" from surface water bodies such as streams and ponds, to replace the water being lost from storage. The sustainable yield of an aquifer must be considerably less than recharge if adequate amounts of water are to be available to sustain both the quantity and quality of streams, springs, wetlands, and groundwater-dependent ecosystems. In reality, when large withdrawals from the aquifer continue to drain water from storage, the aquifer system as a whole shrinks.

When a well field is operated, even if the general input is much greater than extraction, a transient state will always occur before the water levels in wells stabilize. The duration of the transient state depends mainly on aquifer characteristics such as size and hydraulic diffusivity, degree of stratification, and heterogeneity. On the other hand, the natural recharge of an aquifer in semi-arid and arid climates does not have a linear relationship with precipitation. In dry years, recharge might be negligible or even negative due to evapotranspiration or evaporation from the water table. Significant recharge may only occur once every one or more decades. Therefore, the water table depletion trend during a long dry spell – when the recharge is almost nil and the pumping is high – might not be representative of a long-term situation.

Groundwater abstraction from the transboundary Indo-Gangetic Basin comprises 25% of global groundwater withdrawals, sustaining agricultural productivity in India and adjacent countries. Recent interpretations of satellite gravity data indicate that current abstraction is unsustainable, but these large-scale interpretations lack the spatiotemporal resolution required to govern groundwater effectively. The volume of groundwater at the 200 m depth is estimated to be >20 times the combined annual flow of the Indus, Brahmaputra, and Ganges, and the water table was stable or rising across 70% of the aquifers between 2000 and 2012 (MacDonald et al. 2016). Groundwater levels are falling in the remaining 30%, amounting to a net annual depletion of $8.0 \pm 3.0\ km^3$. Recent groundwater depletion in northern India has occurred within a longer history of groundwater accumulation from extensive canal leakage. New evidence is available from high-resolution *in situ* records of groundwater levels, abstraction, and water quality, which reveal that sustainable groundwater supplies are constrained more by extensive contamination than depletion.

3.1.13 Reasons for Groundwater Pollution

As described in Chapter 1, in India, where groundwater is used intensively for irrigation and industrial purposes, a variety of land use and water-based human activities are causing pollution of groundwater. Its overexploitation is causing aquifer contamination in certain instances, while in others its unscientific development with insufficient knowledge of groundwater flow dynamic and geo-hydrochemical processes has led to its mineralization. Intensive use of chemical fertilizers and pesticides in agricultural farms, often dispersed over large areas, and indiscriminate disposal of human and animal waste on land result in leaching of the residual nitrate and other chemicals, causing non-point pollution in groundwater, which is a great threat to fresh groundwater. Soil texture, pattern of fertilizer and pesticide use, their degradation products, and total organic matter in the soil govern the vulnerability of groundwater to pesticide and fertilizer pollution.

The above problems usually affect the short to medium term. Efforts to overcome them should not distract the attention of water policy decision makers from the most serious medium- to long-term problem: groundwater contamination. However, this problem is more linked to land use planning than to the intensive use of groundwater. The degradation of groundwater quality may not be related at all to excessive abstraction of groundwater in relation to average natural recharge. Other causes may be responsible, such as return flow from surface water irrigation, leakage from urban sewers, infiltration ponds for waste waters, septic tanks, urban solid waste landfills, abandoned wells, mine tailings and many other activities not related to groundwater development. Also a temporary situation, such as a serious drought, can contribute to the degradation of groundwater quality.

Irrigated agriculture can contribute to lasting deterioration of groundwater quality, preventing its use for drinking and endangering ecosystems and biodiversity. Highly profitable agriculture comes with a cost, unless best management practices are carefully adhered to and enforced – something that still needs to be recognized in the policy debate. There is a need to move towards more sustainable agricultural water management, marking a useful first step that should now be followed with detailed advocacy, action and implementation plans for more sustainable groundwater use. Feeding people from the ground up will also require changes in the public view of groundwater – not as a magical source of free water from the ground but rather a finite and precious resource that needs to be collectively and effectively managed.

Groundwater abstraction can cause, directly or indirectly, changes in groundwater quality. The intrusion into a fresh-water aquifer of low-quality surface or groundwater because of a change in the hydraulic gradient due to groundwater abstraction is a frequent cause of quality degradation. Saline intrusion may be an important concern for the development of aquifers adjacent to saline water bodies. This is a typical problem in many coastal regions of semi-arid or arid regions (Rina et al. 2012) The relevance of the saline water intrusion not only depends on the amount of abstraction, in relation to the natural groundwater recharge, but also on the well field location and design, and the geometry and hydrogeological parameters of the pumped aquifer. In many cases, the existing problems are due to uncontrolled and unplanned groundwater development and not to excessive pumping.

Public ignorance continues to exist about environmental considerations, provision of basic social services, unplanned agrochemicals application, disposal of improperly treated sewage water and anthropogenic wastes on land, into surface waterbodies and unlined drains, inducing leaching/seepage of contaminants from landfills, canals, rivers, and drains. The distribution of ^{18}O isotope and chemical contaminant species clearly indicates that slow infiltration of agricultural and urban runoff along with pollutants present in excessively applied agrochemicals and resulting wastes were major causes of pollution affecting the water potential (Datta et al. 1996b, 1997).

The contamination level at a specific location depends on evaporation/recharge, adsorption/dispersion in the soil zone, intermixing of polluted groundwater with fresh water along specific flow pathways, and increasing lateral extension of polluted water (Datta and Tyagi 1995; Datta et al. 1996a, 1997; Datta 2013a, 2015). The groundwater flow direction and overexploitation-induced changes in hydraulic head result in mixing of highly saline/contaminated groundwater with relatively fresh groundwater or river water along specific flow pathways, thereby increasing lateral extension of contaminated groundwater and decreasing the fresh-water potential. There are inadequate or absent estimates of the public health consequences of groundwater pollution as it involves methodological complexities and logistical problems. Nevertheless, levels of toxicity depend on the type of pollutant.

Limited efforts have been made to create efficient irrigation systems, water management, and modifications in agronomic practices and soil amendments, along with changes in cropping sequences to reduce the impact of irrigated production on offsite water quantity and quality (Datta et al. 1997; Datta 2005, 2013a, 2015). Better integration of laboratory data management with continuous water quality and quantity monitoring data management is needed to reconcile the problem of how sparse, dense information can better inform watershed-scale objectives. An awareness of the crucial importance of preventing groundwater pollution in order to avoid a future water crisis exists only in a few countries. A strong awareness effort is needed in order not to bequeath future generations some of our better aquifers almost irreversibly polluted. The depletion of groundwater storage is generally a problem as serious as groundwater quality degradation, and can perhaps be solved without great difficulty, for example if water use efficiency is improved.

3.1.14 Groundwater Vulnerability to Floods, Risk Mitigation, and Water Security

Over the decades, worldwide in many riverine and coastal areas including India and South-EastAsia (e.g. Cambodia, Thailand, Vietnam, Laos, Philippines), Australia, the US, Canada, and Indonesia, flood frequency and severity have increased, altering the landscape and storm water channels in rural and urban environments. Both spatial and temporal variations in storms and heavy prolonged rainfall-induced floods mask, regulate or aggravate water crisis in different areas at different times. In most areas, to meet water demand, dependence on groundwater has increased, and overexploitation has continued for drinking and other purposes due to increasing populations, the developmental boom, intensification of urbanization, agricultural, industrial, and commercial activities. With water demand projected to outstrip supply, the greatest risks are likely to be water insecurity, failure of water supply and sewer systems, groundwater pollution, and sea water intrusion in affected areas. In many flood-affected areas drinking water supply is met mainly from aquifers. Flood risk mitigation initiatives for safe water supply often face challenges, even if groundwater is accessible locally, because of limited knowledge about the major sources of groundwater.

3.1.14.1 Knowledge Required for Flood Risk Mitigation

During floods, due to ineffective water supply systems, identification of groundwater sources from deep aquifers or even non-renewable "fossil" water sources with adequate yields in the affected region can be usefully exploited until the regular water supply system is restored. For comprehensive evaluation of risk severity and mitigation at a specific location, emerging issues are management of knowledge, unawareness, and different perceptions of vulnerability and risks among politicians, managers, policy makers, and the public. A holistic approach by timely investigation with cutting-edge research, knowledge, mapping, early warning, etc. can be helpful for identification of such aquifers by detailed assessment of aggregate situations of groundwater occurrence, accessibility, rechargeability, productivity, residence time, recharge zones, water quality, potential under river floodplains, vulnerability, and sea water intrusion. Multi-isotopic approaches have provided detailed insights into the groundwater situation in India and many other parts of the world. These approaches can help in framing proper flood risk mitigation strategies.

3.1.14.1.1 Multi-isotope Approaches Adopted in India

In each area, groundwater recharge has been assessed separately using artificially injected ^{3}H and environmental isotopes (^{2}H, ^{3}H, ^{14}C, ^{18}O) in water cycle components (Datta et al. 1973; Datta and Goel 1977; Goel et al. 1977; Datta et al. 1980a, 1994, 1996a; Datta 2013a, b, 2015). The spatial and temporal variations of $^{18}O/^{16}O$ and $^{2}H/^{1}H$ ratios in rainwater, groundwater, and river water were monitored to provide insight into groundwater provenance, recharge characteristics, flow pathways of intermixing, contamination and dynamics, surface water–groundwater interactions, and potential under river flood plains (Datta et al. 1996b, 1997, 1999, 2001). ^{14}C ($T_{1/2} = 5730$ year) was used to estimate residence time after correcting for ^{14}C in carbonates and bicarbonates in water (Borole et al. 1979). ^{2}H and ^{18}O values are expressed as $\delta^{2}H$ and $\delta^{18}O$ in per mille (‰) deviations from the Standard Mean Ocean Water. Methods of water sampling and isotopic and chemical analyses are described in the cited references.

3.1.14.1.2 Evaluating the Groundwater Recharge and Residence Time

The average groundwater recharge is estimated to be <8–20% of the annual rainfall (~66–100 cm) in flood-prone parts of the Delhi area, Western Uttar Pradesh, Haryana, Punjab, and Gujarat (Datta and Goel 1977; Goel et al. 1977; Datta 2013a, b, 2015; Datta et al. 1973, 1980a, 2001), and 1–14% of annual rainfall (~14–30 cm) in Rajasthan (Sharma and Gupta 1987). In South-East Asia, there is ~8–10 cm long-term average groundwater recharge from ~90–140 cm annual rainfall (Döll et al. 2002). In most places, low recharge suggests limited groundwater renewal potential over recent decades.

On the premise that isotopic composition of groundwater is a proxy indicator of rainfall isotopic composition (unless the water has been subjected to evaporation), let us analyze the isotopic compositions of rainfall and groundwater.

Over much of India, the IAEA-GNIP data show rainfall mean $\delta^{18}O$ (−2 to −6‰), and long-term average Delhi rainfall monthly mean $\delta^{18}O$ (−15.3‰ to +8.0‰) and $\delta^{2}H$ (−120‰ to +55.0‰). Depleted $\delta^{18}O$ is associated with heavy rainfall which causes flooding, and the $\delta^{18}O$ temporal variability depends on rainfall intensity, and the movement of moist air masses from which rain occurs (Datta et al. 1991). Rainfall-deficient years have relatively $\delta^{18}O$ -enriched rainfall compared to normal monsoon years' monthly rainfall $\delta^{18}O$. Borneo rainfall three-year time-series also indicate similar pattern with depleted $\delta^{18}O$ (−10‰) during late boreal summer and relatively enriched $\delta^{18}O$ (−4‰) in late boreal winter, and the seasonal cycle of rainfall $\delta^{18}O$ depends on seasonally varying moisture trajectories (Cobb et al. 2007).

Sumatra, Indonesia, showed daily rainfall $\delta^{18}O$ −15 to 0‰, with 50% precipitable water of Indian Ocean origin. The groundwater $\delta^{18}O$ data of India indicate <−4‰ (NW and Gangetic Plains), −4 to −2‰ (SE coast Plains) and >−2‰ (Western Ghats and Deccan Plateau) (Datta 2013a, b, 2015), with significant evaporative enrichment of rainfall isotopes before recharge in most areas, particularly in western, central and eastern parts and the eastern side of the Western Ghats. In south India, NE monsoon-dominated groundwater has depleted ^{18}O and ^{2}H compositions compared to SW monsoon-contributed groundwater. In west and east coastal areas, rainfall and groundwater isotopic content has reasonable similarity.

In flood-prone NW and Gangetic Plains, where the SW monsoon generates large amounts of runoff, highly depleted $\delta^{18}O$ (−4.2 to −7.6‰) in groundwater compares to the current weighted mean rainfall $\delta^{18}O$, with a ^{3}H age of <50 years and ^{14}C age of 2000–22 000 years BP, in Gujarat and Rajasthan (Borole et al. 1979; Datta 2013a, b, 2015), suggesting groundwater $\delta^{18}O$ is mixture of the long-term weighted average rainfall and the "flood-generating rainfall," and slow flushing groundwater in the last interglacial period with a dry-arid climate. While ^{3}H presence in groundwater suggests rapid circulating recent recharge, absence of ^{3}H indicates slow-moving/trapped groundwater with transient time >50 years.

In the Barmer, Phalodi, and Jaore areas of Rajasthan, deeper and artesian groundwater with negligible/low ^{3}H with a maximum ^{14}C age of 4000 years suggests that paleo-waters were recharged during a more humid period. In the Hyderabad urban area, granitic aquifers with various sets of fractures/joints contain deep groundwaters with similar $\delta^{18}O$ (−3.2 to −1.7‰) and ^{14}C age <1000–5600 years BP, suggesting the same source of recharge but different transit paths (Kumar et al. 2008). In Malaysia (Mostapa et al. 2011) and the Bangkok Basin, Thailand (Sanford 1997), there are reports of modern recharge in shallow aquifers, and in some places paleo-water has been recharged during the past relatively humid climate. Groundwater in old marine formations shows high ^{14}C age but young ^{3}H age.

3.2 Identifying Groundwater Recharge Zones

Generally, in basin structures with deep aquifers, groundwater with longer residence time has low vulnerability (Datta 2013a, b, 2015). In the active recharge zones, groundwater is young (<50 years) and quite susceptible to contaminants. In the passive recharge zones, groundwater is older (>100 years) and better protected against pollutants, but could be more mineralized than in the active recharge zone, and sometimes may have dissolved elements (e.g. As, F), because of high mean turnover times. Based on ^{3}H ages, the Delhi area (located in the flood-prone Gangetic Plains) can be grouped into three recharge zones.

- In the highland area, modern recharge (<8 years old; $^{3}H > 10$ TU), extending to intermediate (20–30 m) to deeper depths, in the eastern Yamuna floodplain at deeper depths (>30 m) and in some western upstream and downstream sites, and in the vicinity of the Western Yamuna Canal and drains at shallow (<20 m) to intermediate depths.
- In most other parts, submodern water with ^{3}H 10–5 TU; age 8–20 years.
- Old water with ^{3}H <5 TU; age > 20 years (Kumar et al. 2011; Datta 2013a).

Moving away from canals/drains, a gradual increase in groundwater age (decrease in ^{3}H) suggests a different degree of mixing of old groundwater with young surface water. In the Delhi area, Pushkar Valley and Jaisalmer District, Rajasthan, and Sabarmati Basin, Gujarat, groundwater ^{14}C age and $\delta^{18}O$ depth variation suggest vertical stratification (Datta et al. 1980a, 1994; Datta 2013b).

Recharge zones of springs (altitude 910–1330 m above mean sea level (amsl)) were identified in mountainous regions of Chamoli district, Uttarakhand, using ^{3}H, $\delta^{18}O$ and δD (Kumar et al. 2008). Rain from 880, 990 and 1180 m amsl indicated $\delta^{18}O$ depletion −0.55‰ per 100 m increase in altitude and −3.8‰ for δD, suggested that the high-altitude springs were provoked by recent rainfall recharge (^{3}H ~11.5 TU), while the lower altitude springs were old (^{3}H 9.5 TU) and subjected to evaporation ($\delta^{18}O$ enriched by 2‰) prior to recharge. The recharge area of lower altitude springs was assessed to be 1250 m amsl (close to 1180 m amsl high-altitude springs). In parts of Haridwar and Saharanpur districts of the Siwalik foothills in India, based on high ^{3}H (>9 TU) in the deep aquifer groundwater from depths >60 m, two recharge zones were identified along the Upper Ganga Canal and Eastern Yamuna Canal plains, and one in the area above the plains (Kumar et al. 2008). A comparison of the $\delta^{18}O$ in groundwater, rainfall, and canal, with 0.31‰ depletion in $\delta^{18}O$ per 100 m increase in altitude, suggests that for the recharge area above the plains (>550 m amsl), the groundwater $\delta^{18}O$ (−18.9 to −8‰) is similar to $\delta^{18}O$ in rainfall over an altitude of 600–1000 m amsl in the Bhabhar-Siwalik region.

3.2.1 Induced Aquifer Recharge to Natural Recharge, and Flooding Impacts

For an aquifer in hydrological continuity with a river/sea region, determining groundwater recharge, $\delta^{18}O$ distributions, and GW-SW effluent/influent seepage helped to identify the intermixing pathways and bridge gaps in groundwater management under floodplains. Seasonal variations in $\delta^{18}O$ composition of the Yamuna and its tributaries were associated with most depleted values during the monsoon period with high rainfall. To conserve floodwater under the Yamuna river floodplains, a simple conceptual mixing model using the river water $\delta^{18}O$ (−9.7‰) and adjacent groundwater $\delta^{18}O$ (−5.6 to −9.6‰) in shallow wells indicated good interconnection of river flow and the adjacent groundwater, and 2–96% of the river water contributed to groundwater under the floodplains at different stretches (Datta and Kumar 2011; Datta 2013a, 2015). To make recharge more responsive to flooding, large volumes of runoff can be conserved under floodplains via the hyporheic zone, and the high recharge potential in areas close to river floodplains and canals should be protected.

In Barmer and Jalore districts in western Rajasthan (where relatively higher rainfall occurs), in the shallow (away from river course), deep and artesian groundwater, depleted $\delta^{18}O$ (−3.7 to −7.3‰) and $\delta^{2}H$ (−24.8 to −55.4‰) suggest interconnection of shallow and deep aquifers (Kumar et al. 2008). However, shallow groundwater along a river course with high ^{3}H (5–20 TU) indicates modern recharge from rainfall and through river channels during episodic floods. Wells away from the river with depleted ^{3}H (1–4 TU) indicate less permeable zones for vertical recharge. In the shallow groundwater (<50 m) away from the River Sukri (a tributary of the River Luni) and deep groundwater along the river, depleted $\delta^{18}O$ (−4 to −7‰) with low ^{3}H content (1–4.8 TU) and higher chloride content (400–1000 ppm) suggests that intense rain-induced flash flood events freshen brackish groundwater by river water recharge (Kumar et al. 2008). In the Bhabhar-Siwalik plains, groundwater with $\delta^{18}O$ (−7 to −9‰) in the recharge zones and $\delta^{18}O$-depleted groundwater within 1.5 km of a canal suggest groundwater recharge through canal seepage. The $\delta^{18}O$ enrichment in groundwater with distance away from canal suggests mixing with water recharged from rainfall.

3.2.2 Monitoring Groundwater Vulnerability to Pollution and Risk Mitigation

In India and many South-East Asian cities, aquifers contaminated with increased levels of chloride and TDS are a serious concern. To mitigate flood risk, while structural measures may have adverse effects on the ecosystem, non-structural measures (activating deep wells, springs, etc.) can be useful to check groundwater vulnerability to pollution. The isotopic approaches (described below) integrated with hydrochemistry can help to determine groundwater source, mixing, travel time, pollutant level and pathways from the source to the water supply wells, temporal leaching characteristics, and an aquifer's attenuation capacity for pollutants under natural and exploited conditions. The groundwater recharge zones need to be clearly identified and revised in relation to land use changes, and unplanned agro chemicals application and waste disposal in such areas should be restricted or eliminated.

In north-west India, in low permeability deeper aquifers and in some shallow aquifers, high salinity groundwater is preserved by direct interaction with lakes and rivers, and indirectly by rainfall recharge during the Last Glacial Maximum (LGM) (30 000–12 000 years BP) (Singh et al. 1974; Datta 2013a, 2015). In flood-affected parts of India, fresh-water availability is constrained by both past and ongoing pollution of groundwater with moderate to high salinity, F (<1–46.0 mg L^{-1}), NO_3 (<20–1600 mg L^{-1}), Zn, Cu, Fe, Mn, Pb, Ni, Cd, etc. exceeding the WHO MPL in drinking water, contributed from both non-point and point sources (Datta 2013a, 2015; Datta et al. 1996b, 1997, 1999,

2001). The straight-line relationships of ^{18}O and pollutant species and the iso-contours of contaminant spatial distribution helped in identifying the polluted groundwater plume direction and flow pathways (Datta et al. 1996a, b, 1997; Datta 2013a, 2015). Overexploitation-induced lateral intermixing of polluted groundwater or river water (in floodplains) with fresh water along specific flow paths increases lateral extension of polluted groundwater, decreasing the fresh-water potential. Groundwater ^{18}O iso-contour small gradients indicate hydrodynamic zones (Datta et al. 1994; Datta 2013a, b, 2015).

In the Delhi area, scatter plots of electrical conductivity, $\delta^{18}O$ and ^{3}H content suggest that the modern groundwater is fresh and the older groundwater is saline, and two main flow systems exist vertically: (i) uppermost rapidly circulating, low salinity local flow – more vulnerable to overexploitation; and (ii) relatively slow circulating intermediate zone – more vulnerable to salinity and depletion. In the Pushkar (Ajmer) valley, Rajasthan, $\delta^{18}O$ and hydrochemistry helped in assessing the possible existence of both fossil and replenishable groundwater, and causes of salinity (Datta et al. 1994). The $\delta^{18}O$ of groundwater ranged from – 6.3 to +4‰, but most of the groundwater had a more depleted $\delta^{18}O$ than the rainfall annual average $\delta^{18}O$ (−2.5‰) in Rajasthan. The highly depleted ^{18}O and low salinity groundwater could be of paleo-origin at deeper depths; intermediate ^{18}O values with high salinity in aquifers at intermediate depths are the result of evaporative enrichment during the infiltration of runoff and dissolution of salts in an evaporatic environment, and the shallow groundwater has highly enriched ^{18}O and high salinity. In the Delhi area and Sabarmati Basin, a highly skewed and wide range of groundwater Cl, NO_3, F, and heavy metals associated with enriched ^{18}O suggests pollution by infiltration of evaporated surface runoff and rainwater along with pollutants from both point and non-point sources (Datta et al. 1996a, b, 1998; Rina et al. 2011, 2013).

3.2.3 Sea Water Intrusion

In coastal areas of Gujarat, Cl concentration and Na/Cl ratio suggest impacts of seawater relics and intrusion on the groundwater; there is occurrence of fresh water (Cl^- <250 mg L^{-1}), and saline water (Cl^- >250 mg L^{-1} and Na/Cl ≤1), and the salinity front movement from the coast up to 25, 30, and 50 km in shallow, intermediate, and deeper aquifers respectively (Rina et al. 2011, 2013). In a coastal aquifer in the Krishna River Delta, India, it was possible to identify recharge sources/zones and sea water intrusion. Measurements of ^{18}O, ^{2}H, ^{3}H and ^{14}C from shallow (<30 m), intermediate (30–60 m) and deep (>60 m) groundwater and water from the River Krishna and its tributaries, the Bay of Bengal and Prakasam Reservoir suggest that the Prakasam Barrage and canal networks are the major sources of recharge, and contribute significantly to freshen the saline groundwater (Kumar et al. 2008). Current sea water intrusion is limited to only a few locations in shallow, intermediate, and deep aquifers up to 27, 32, and 50 km distance from the coast respectively. In shallow and intermediate groundwater, low ^{3}H (0–4 TU) suggests that recharge at thePrakasam Barrage area is via the canal network, particularly in the area near the eastern delta and a few paleo-channels (on which canals are laid). Decreasing ^{3}H content with water passage to the coast suggests old water dilution by mixing with modern water, and mixing efficiency decreases with depth. The uncorrected ^{14}C age of 2500–37 000 year BP, with an increasing trend towards the coast in deeper groundwater with low or negligible ^3H, suggests a paleo-marine origin of deep, old, and saline groundwater.

3.2.3.1 Natural Recharge of Coastal Aquifers

Many coastal zones around the world have irregular precipitation throughout the year. This results in discontinuous natural recharge of coastal aquifers, which affects the size of freshwater lenses present in sandy deposits. The influence of variable/discontinuous recharge on the size of freshwater lenses was simulated with the SEAWAT model by defining potential recharge (precipitation minus potential evapotranspiration) at Ameland (The Netherlands), Auckland and Wellington (New Zealand), Hong Kong (China), Ravenna (Italy), Mekong (Vietnam), Mumbai (India), New Jersey (USA), Nile delta (Egypt), Kobe and Tokyo (Japan), and Singapore (Mollema and Antonellini 2013). The discrepancy in models with continuous and discontinuous recharge is relatively small in areas with low total annual recharge (258–616 mm $year^{-1}$) but in places with a monsoon climate (e.g. Mumbai, with recharge up to 1686 mm $year^{-1}$), the difference in fresh-water lens thickness between the discontinuous and the continuous model is larger (up to 5 m) and thus important to consider in numerical models that estimate fresh-water availability.

3.2.4 Monitoring the Groundwater Situation in Africa

In rural Africa, most drinking water demands are met from shallow groundwater (<50 mbgl) but there is little information on the resilience of this resource to future changes. In West Africa, to estimate the residence times of shallow groundwater in sedimentary and basement aquifers, and investigate the relationship between groundwater resources and climate, stable isotopes, CFCs, SF_6, and ^3H, were used in different climatic zones with annual rainfall 400–2000 mm $year^{-1}$ (Lapworth et al. 2013). Stable isotope results indicate that most of the shallow groundwater is recharged rapidly following rainfall. Most of the groundwater has a mean residence times of 32–65 years. Chloride mass balance results indicate that in the arid areas (<400 mm annual rainfall), there is recharge of up to 20 mm $year^{-1}$. Similar residence times in both the sedimentary and basement aquifers suggest uniform hydraulic diffusivity and significant groundwater storage within the shallow basement. This further suggests considerable resilience to short-term interannual variation in rainfall and recharge, and rural groundwater resources are likely to get adjusted to diffuse, low-volume abstraction.

Although climate change is likely to have an impact on future recharge rates and hence on the underlying groundwater resources, for appropriate validation of the simulation of present and future scenarios, more field data and information are needed. The impact may not necessarily be a negative one, as evidenced by some of the investigations. Quantifying the climate impact is difficult and subject to uncertainties present in future climate predictions. Simulations based on general circulation models (GCMs) have yielded mixed and conflicting results, raising questions about their reliability in predicting future hydrological conditions. It has to be kept in mind that ascertaining the usefulness of rainwater harvesting under the anticipated global warming threat is complex, and the consequences of climate phenomena for flood risks are many.

Many of the world's river basins are "closed," with water use equal to or exceeding available renewable water resources, and others are approaching this state. Some countries are contemplating long-distance water transfers to offset local water shortages for multiple needs, while simultaneously aiming to increase water use efficiency and reduce demand. The rate of water storage creation in many developing countries in Africa and Asia is considered inadequate, however, to meet future challenges related to climate change and rising demand for water from other sectors. A lack of water storage infrastructure may result in heavy economic losses due to flooding and drought and impose high costs on human health from polluted water.

Groundwater use in agriculture has accelerated with the advent of motorized pumps. Irrigated areas reliant on stored groundwater are estimated to occupy almost 40% of the global area equipped for irrigation. This demand for groundwater has helped intensify agricultural production, both in existing surface irrigation commands and on land that would otherwise have no supply. The pressure on aquifers is such that many important groundwater systems are now overabstracted, and the associated loss of water quality is increasing. Moreover, groundwater abstraction with motorized pumps uses large amounts of energy, which is expensive for farmers, and if fossil fuels are used, this influences climate change. Given the anticipated impacts of climate change on long-term recharge patterns, in many situations, the overexploitation of groundwater can be halted with the development of additional storage capacity that reduces the burden on aquifers and by encouraging farmers and other water users to keep to sustainable abstraction levels.

For effective flood risk mitigation and to implement measures for better preparedness and strategies for water security, it is desirable not only to create physical infrastructure but also to analyze flood frequency, magnitude, and possible causes. It is also necessary to remap the changes in floodplain topography, streams and river flow paths, to evaluate issues emerging from previous events and the economic, social, physical or geographic factors that reduce public resilience to cope with the impact of potential flooding. Integrating these information and isotopic approaches can generate knowledge on vulnerable areas and flood risk by regular monitoring of groundwater occurrence, dynamics, and recharge, delineating potential recharge, quality and protection zones which differ from region to region and within a region, and restricting waste disposal in these zones. Assessing spatiotemporal variation in different time scales of groundwater recharge can help to create an integrated system of adequate water supply.

In climate change impact studies, precipitation is usually downscaled and hence the result is often poor or unreliable. A poor correlation is often attributed to meso-scale precipitation processes generally occurring in the summer season at the site scale that are not represented in regional models due to their spatial and temporal sizes in comparison to larger scale regional precipitation. As a result, global-scale models are likely to underestimate local summer precipitation. Moreover, research on the impacts of climate change on the groundwater system is relatively limited due to the lack of long-term historical data. Also, the driving forces that cause such changes are still unclear. A model of groundwater systems under possible climate change based on available data is very important to prevent the deterioration of regional water resource problems in the future.

Investigation of the relationship between climate change and loss of fresh groundwater resources is important for understanding the characteristics of different regions. Groundwater recharge is influenced not only by hydrological processes but also by the physical characteristics of the land surface and soil profile. Many climate change studies have focused on modeling the temporal changes in the hydrological processes and ignored the spatial variability of physical properties across the study area. While knowing the average change in recharge and groundwater levels over time is important, these changes will not occur equally over a regional catchment or watershed. Long-term water resource planning requires both spatial and temporal information on groundwater recharge in order to properly manage not only water use and exploitation, but also land use allocation and development. Studies concerned with climate change should therefore also consider the spatial change in groundwater recharge rates.

By integrating water harvesting and storage structures in balance with groundwater recharge and abstraction in a planned and systematic manner, it may be possible to create a "water buffer," which can help reduce vulnerability to drought and seasonal variations in rainfall and can be used for multiple purposes, such as agriculture, livestock watering, and domestic use. India and many other countries have a long tradition of efficient systems for water harvesting or tapping springs, developed from ancient times in some of the most arid and water-stressed regions with different terrains and climatic conditions. In most cases, the main purpose of rainwater harvesting systems is not only to make water available in ponds or storage tanks, but also to allow water to percolate down into the ground to recharge the groundwater.

Although governments and NGOs have shown interest in rainwater harvesting, there is a limited scientific knowledge base for systematically putting RWH into practice for artificial recharging of groundwater. If the likely consequences of future changes in groundwater recharge, resulting from both climate and socioeconomic change, are to be assessed, hydrogeologists must be increasingly consulted by the NGOs and government to work with researchers from other disciplines, such as socioeconomists, agricultural modelers, and soil scientists. It is essential to create awareness among politicians, the media, and public to change mindsets and behavior, develop transparency, share knowledge, trust, interests and responsibility, and properly employ resources, guided by ethical considerations.

References

Athavale, R.N., Rangarajan, R., and Muralidharan, D. (1992). Measurement of natural recharge in India. *Journal of the Geological Society of India* 39 (3): 235–244.

Bhanja, S.N., Abhijit, M., Matthew, R. et al. (2017). Groundwater rejuvenation in parts of India influenced by water-policy change implementation. *Scientific Reports* 7: 7453.

Borole, D.V., Gupta, S.K., Krishanswami, S. et al. (1979). Uranium isotopic investigations and radiocarbon measurement of river-groundwater systems, Sabarmati Basin, Gujarat, India. In: *Isotope Hydrology*, vol. 1, IAEA-SM-228/11, 118–201. Vienna: IAEA.

CGWB (Central Ground Water Board) (1998). Groundwater Resources of India. Ministry of Water Resources, New Delhi.

CGWB (Central Ground Water Board) (2014–2015). *Year Book*. New Delhi: Ministry of Water Resources.

Chatterjee, R. and Purohit, R.R. (2009). Estimation of replenishable groundwater resources of India and their status of utilization. *Current Science* 96: 1581–1591.

Cobb, K.M., Adkins, J.F., Partin, J.W., and Brian, C. (2007). Regional-scale climate influences on temporal variations of rainwater and cave drip water oxygen isotopes in northern Borneo. *Earth and Planetary Science Letters* 263: 207–220.

Datta, P.S. (2005). Groundwater ethics for its sustainability. *Current Science* 89 (5): 812–817.
Datta, P.S. (2008). *Water: A Key Driving Force*. New Delhi: Vigyan Prasar Publication.
Datta P.S. (2013a). Groundwater vulnerability to changes in land use and society in India. In: Understanding Freshwater Quality Problems in a Changing World, Proceedings of H04, IAHS-IAPSO-IASPEI Assembly, Gothenburg, July (IAHS Publ. 361, 2013), pp. 345–352.
Datta P.S. (2013b) Geoethics and geoscientists role to protect geodiversity for societal development GEOITALIA2013, Session Geopolicy – L 2 Geoethics and Society: Geosciences Serving the Public, September 16–18, Pisa.
Datta P.S. (2014). Need for better groundwater governance for sustained water supply. IPHE National Conference on Piped Water Supply and Sewerage Systems, January 24–25, New Delhi.
Datta, P.S. (2015). Ethics to protect groundwater from depletion in India. In: *Geoethics: The Role and Responsibility of Geoscientists* (ed. S. Peppoloni and G. di Capua), 19–24. London, Special Publication no. 419: Geological Society.
Datta, P.S. (2017). *Better Groundwater Governance Only Can Ensure Sustained Water Supply*. Germany: Lambert Academic Publishing.
Datta, P.S., Bhattacharya, S.K., Mookerjee, P., and Tyagi, S.K. (1994). Study of groundwater occurence and mixing in Pushkar (Ajmer) Valley, Rajasthan with ^{18}O and hydrochemical data. *Journal of the Geological Society of India* 43: 446–456.
Datta, P.S., Bhattacharya, S.K., and Tyagi, S.K. (1996a). ^{18}O studies on recharge of phreatic aquifers and groundwater flow-paths of mixing in Delhi area. *Journal of Hydrology* 176: 25–36.
Datta, P.S., Deb, D.L., and Tyagi, S.K. (1996b). Stable isotope (^{18}O) investigations on the processes controlling fluoride contamination of groundwater. *Journal of Contaminant Hydrology* 24 (1): 85–96.
Datta, P.S., Deb, D.L., and Tyagi, S.K. (1997). Assessment of groundwater contamination from fertilizers in Delhi area based on ^{18}O, NO_3^- and K^+ composition. *Journal of Contaminant Hydrology* 27 (3–4): 249–262.
Datta, P.S., Desai, B.I., and Gupta, S.K. (1979). Comparative study of groundwater recharge rates in parts of Indo-Gangetic and Sabarmati alluvium plains. *Mausam* 30 (1): 129–133.
Datta, P.S., Desai, B.I., and Gupta, S.K. (1980a). Hydrological investigations in Sabarmati basin-I. groundwater recharge estimation using tritium tagging method. *Proceedings of the Indian National Science Academy, Part A: Physical Sciences* 46 (1): 84–98.
Datta, P.S. and Goel, P.S. (1977). Groundwater recharge in Punjab state (India) using tritium tracer. *Nordic Hydrology* 8: 225–236.
Datta, P.S., Goel, P.S., Rama, and Sangal, S.P. (1973). Groundwater recharge in western Utter Pradesh. *Proceedings in Mathematical Sciences* 78: 1–12.
Datta, P.S., Gupta, S.K., and Sharma, S.C. (1980b). A conceptual model of water transport through the unsaturated soil zone. *Mausam* 31 (1): 9–18.
Datta, P.S. and Kumar, S. (2011) Significance of ^{18}O and hydrochemical composition to characterize water dynamics in hyporheic zone of Yamuna river flood plains in Delhi Area. Proceedings of the 3rd International Multi-disciplinary Conference on Hydrology and Ecology: Ecosystems, Groundwater and Surface Water – Pressures and Options, May 2–5, Austria.
Datta, P.S., Manjaiah, K.M. & Tyagi, S.K. (1999) ^{18}O isotopic characterisation of non-point source contributed heavy metal (Zn and Cu) contamination of groundwater. Proceedings of International Symposium on Isotope Techniques in Water Resources Development and Management, May 10–14, Vienna, pp. 190–198.
Datta P.S., Rohilla S.K., and Tyagi S.K. (2001). Integrated approach for water resources management in Delhi region: problems and perspectives. In: Regional Management of Water Resources, Proceedings of Maastricht Assembly Symposium, July (IAHS Publ. No. 268), pp. 1–8.
Datta, P.S. and Tyagi, S.K. (1995). Groundwater inter mixing model and recharge conditions in Delhi area as derived from oxygen-18 and Deuterium. In: *Sub-Surface Water Hydrology*, 103–119. Netherlands: Kluwer Academic.
Datta, P.S., Tyagi, K., and Chandrasekharan, H. (1991). Factors controlling stable isotopes in precipitation in New Delhi, India. *Journal of Hydrology* 128: 223–236.
Datta, P.S., Tyagi, S.K., Mookerjee, P. et al. (1998). Groundwater NO_3 and F contamination processes in Pushkar Valley, Rajasthan as reflected from ^{18}O isotopic signature and ^3H recharge studies. *Environmental Monitoring and Assessment* 56 (2): 209–219.

Döll, P., Lehner, B., and Kaspar, F. (2002). Global modeling of groundwater recharge. In: *Proceedings of Third International Conference on Water Resources and the Environment Research*, vol. I (ed. G.H. Schmitz), 27–31. Germany: Technical University of Dresden.

Goel, P.S., Datta, P.S., and Tanwar, B.S. (1977). Measurement of vertical recharge to groundwater in Haryana state (India) using tritium tracer. *Nordic Hydrology* 8: 211–224.

Jasechko, S., Perrone, D., Befus, K.M. et al. (2017). Global aquifers dominated by fossil groundwaters but wells vulnerable to modern contamination. *Nature Geoscience* 10: 425–429.

Kulkarni, K.M., Rao, S.M., Singhal, B.B.S., Parkash, B., Navada, S.V., Nair, A.R. (1989). Origin of saline groundwaters of Haryana State, India. Regional Characterization of Water Quality, Proceedings of Baltimore Symposium, May.

Kumar, M., Rao, M.S., Kumar, B., and Ramanathan, A. (2011). Identification of aquifer recharge zones and sources in an urban development area (Delhi, India), by correlating isotopic tracers with hydrological features. *Hydrogeology Journal* 19: 463–474.

Kumar, B., Singh, U.K., and Rao, M.S. (2008). Isotope hydrology in India. *Hydrology Journal* 31 (34): 45–79.

Lapworth, D.J., MacDonald, A.M., Tijani, M.N. et al. (2013). Residence times of shallow groundwater in West Africa: implications for hydrogeology and resilience to future changes in climate. *Hydrogeology* 21 (3): 673–686.

MacDonald, A.M., Bonsor, H.C., Ahmed, K.M. et al. (2016). Groundwater quality and depletion in the Indo-Gangetic Basin mapped from *in situ* observations. *Nature Geoscience* 9: 762–766.

Mollema, P.N. and Antonellini, M. (2013). Seasonal variation in natural recharge of coastal aquifers. *Hydrogeology* 21 (4): 787–797.

Mostapa, R., Mohd-Tadza, S.A.–.R., and Abustan, I. (2011). Comparison between conventional and stable isotope techniques in determining distribution of landfill leachate in groundwater and surface waters in Perak, Malaysia – techniques of water resources investigation. *International Journal of Environmental Sciences* 1 (5): 948–958.

Planning Commission (2007a). Report on Steering Committee on Water Resources for 11th Five Year Plan. Government of India, New Delhi.

Planning Commission (2007b) Report of the Expert Group on Water Management and Ownership. Government of India, New Delhi. http://planningcommission.nic.in/reports/genrep/rep_grndwat.pdf.

Rangarajan, R. and Athawale, R.N. (2000). Annual replenishable ground water potential of India – an estimate based on injected tritium studies. *Journal of Hydrology* 234: 38–53.

Rao, S.M. (2003). Use of isotopes in search of lost river. *Journal of Radioanalytical and Nuclear Chemistry* 257 (1): 5–9.

Rina, K., Datta, P.S., Singh, C.K., and Mukherjee, S. (2011). Characterization and evaluation of processes governing the groundwater quality in parts of the Sabarmati basin, Gujarat using hydrochemistry integrated with GIS. *Hydrological Processes* 26 (10): 1538–1551.

Rina, K., Singh, C.K., Datta, P.S. et al. (2012). Geochemical modelling, ionic ratio and GIS based mapping of groundwater salinity and assessment of governing processes in Northern Gujarat, India. *Environmental Earth Sciences* 69 (7): 2377–2391.

Rina, K., Datta, P.S., Singh, C.K., and Mukherjee, S. (2013). Isotopes and ion chemistry to identify salinization of coastal aquifers of Sabarmati River Basin. *Current Science* 104 (3): 335–344.

Rodell, M., Velicogna, I., and Famiglietti, J.S. (2009). Satellite based estimates of groundwater depletion in India. *Nature* 460: 999–1002.

Sanford, W.E. (1997). Correcting for diffusion in carbon-14 dating of groundwater. *Groundwater* 35 (2): 357–361.

Sharma, P. and Gupta, S.K. (1987). Isotopic investigation of soil water movement: a case study in the Thar Desert, western Rajasthan. *Hydrological Sciences Journal* 32 (4): 469–483.

Singh, G., Joshi, R.D., Chopra, S.K., and Singh, A.B. (1974). Late quaternary history of vegetation and climate of the Rajasthan Desert, India. *Philosophical Transactions. Royal Society of London* 267 (889): 467–501.

Sinha, U.K. and Navada, S.V. (2008). Application of isotope techniques in groundwater recharge studies in arid western Rajasthan, India: some case studies. *Geological Society, London, Special Publications* 288: 121–135.

Soni, V., Gosain, A.K., Datta, P.S., and Singh, D. (2009a). A new scheme for large scale natural water storage in the floodplains: the Delhi Yamuna floodplains as a case study. *Current Science* 96 (10): 1338–1342.

Soni, V., Mehrotra, R., Datta, P.S., and Chander, S. (2009b). A process for organic water. *Current Science* 96 (8): 1100–1103.

Further Reading

Cai, Y., An, Z., Cheng, H. et al. (2006). High-resolution absolute-dated Indian Monsoon record between 53 and 36 ka from Xiaobailong Cave, southwestern China. *Geology* 34: 621–624.

Centre for Science and Environment (2011). *Annual Report 2011-2012*. New Delhi: Centre for Science and Environment.

CGWB (Central Ground Water Board) (1994). *Manual on Artificial Recharge of Groundwater, Technical Series: Monograph No*, vol. 3. New Delhi: Ministry of Water Resources.

CGWB (Central Ground Water Board) (1995). *Report on Groundwater Resources of India*. New Delhi: Ministry of Water Resources.

CGWB (Central Ground Water Board) (2011). *Select Case Studies Rain Harvesting and Artificial Recharge*, 10–11. New Delhi: Ministry of Water Resources.

Datta, P.S. (2016). Ensure smart water availability by behavioral change. In: *Proceedings of the 10th World Aqua Congress, Water-Smart Solutions for Growing India, November 24-25*. New: Delhi.

Gupta, S.K. and Deshpande, R.D. (2005). Water for India in 2050: first-order assessment of available options. *Current Science* 86 (9): 1216–1224.

Karanth, K.R. and Prasad, P.S. (1979). Some studies on hydrologic parameters of groundwater recharge in Andhra Pradesh. *Journal of the Geological Society of India* 20: 404–414.

Kumar, C.P. (2016). Assessing the impact of climate change on groundwater resources. *IWRA (India) Journal* 5 (1): 3–11.

Luthi, D., Le Floch, M., Bereiter, B. et al. (2008). High-resolution carbon dioxide concentration record 650,000-800,000 years before present. *Nature* 453: 379–382.

Tyagi, S.K., Datta, P.S., and Singh, R. (2012). Need for proper water management for food security. *Current Science* 102 (5): 690–695.

4

Phases of Water Harvesting Systems

CHAPTER MENU

4.1 History of Water Harvesting Practices

Rainwater harvesting and water utilization systems have been used since 2000 BCE or maybe earlier. Evidence exists of roof catchment systems in Roman villas to store rainwater as the main water source for drinking and domestic purposes. In India, rainwater harvesting is an ancient technique dating back some 4000–5000 years. Water harvesting has been practiced in India, the Middle East, the Americas, and Africa throughout history and was the backbone of agriculture, especially in arid and semi-arid areas, but the traditional designs varied from region to region governed by several factors (mainly the pattern of rainfall). Ancient religious texts and folk tales give a good insight into the water storage and conservation systems that prevailed in those days. In the Negev desert in Israel, tanks for storing runoff from hillsides in areas with little rain (100 mm year^{-1}) provided water for habitation, domestic, and agricultural purposes. Some of the very earliest agriculture, in the Middle East, was based on techniques such as diversion of wadi flow onto agricultural fields.

The earliest known evidence of the use of the technology in Africa comes from northern Egypt, where tanks ranging in size from 200 to 2000 m^3 have been used for at least 2000 years, and many are still functional today. Rainwater harvesting also has a long history in Asia, where rainwater collection practices have been traced back almost 2000 years in Thailand. The small-scale collection of rainwater from the eaves of roofs or via simple gutters into traditional jars and pots has been practiced in Africa and Asia for thousands of years. In many remote rural areas, this is still used today. The worlds largest rainwater tank is probably the Yerebatan Sarayi in Istanbul, Turkey, constructed during the rule of Justinian (527–565 CE). It measures 140 m by 70 m and has a capacity of 80 000 m^3.

Over the years, with rising populations, growing industrialization, and expanding agriculture, efforts have been made to collect rainwater by building dams and reservoirs

Water Harvesting for Groundwater Management: Issues, Perspectives, Scope, and Challenges, First Edition. Partha Sarathi Datta.

and digging wells; some countries have also tried to recycle and desalinate water. In North America, the agriculture of many indigenous peoples in the present-day southern states was historically dependent on simple methods of floodwater harvesting. In the early twentieth century, the primary focus of conservation agencies was soil erosion control aimed at reducing soil losses; this progressed to soil and water conservation, based particularly on structural measures (terraces, gabion weirs, etc.). The harvesting of runoff that went with some soil conservation measures was more or less a side effect whose potential was unappreciated. Furthermore, the success of the Green Revolution, based on hybrid seeds, inorganic fertilizers and pesticides, resulted in rapid expansion of irrigated areas, and this was seen as the "modern" way forward to improve agricultural water management.

However, since time immemorial around the world, people have known not only how to harvest water but also how to abstract a substantial part of their water need from groundwater. Groundwater withdrawal for the purposes of agricultural expansion soon reached its limits due to overabstraction, declining water resources, and salination, which led to further impoverishment and in some situations to conflicts. Furthermore, the ecological problems associated with dam building became barriers to new construction. Water scarcity and widespread droughts in Africa led to a growing awareness of the potential of water harvesting for improved crop production in the 1970s. After a quieter period in the late 1980s, water harvesting again became the subject of study and project implementation at the turn of the century, and indigenous practices regained credence. In China, water harvesting is seen as a major component in reducing the rural exodus and controlling severe soil erosion and is the subject of dedicated projects, aimed at helping millions of people.

4.2 Phases of Water Harvesting Systems in India

Rainwater harvesting systems (RWHS) provide crucial supplementary irrigation to farmers during water scarcity and dry periods. Small marginal farmers who rely on rain-fed agriculture own the majority of land, and need methods to harvest rain. In most cases, the main purpose of RWHS is not to make water available in ponds or storage tanks, but to allow water to percolate down into the ground to recharge groundwater which can be pumped when needed. In a broad sense, artificial recharging is planned human work to augment groundwater through percolation of surface waters, and for many other purposes, such as conservation of surface runoff and flood waters, control of salt water intrusion, storage of water, prevention of soil erosion and groundwater quality improvement by dilution (Asano 1985). India has a long tradition of water harvesting systems, developed from ancient times in urban and rural areas of some of the most arid and water-stressed hard rock, semi-arid regions. Almost all the forts and most of the old temples in India, in different terrains and climatic conditions, have efficient systems for harvesting rainwater or tapping springs.

4.2.1 Phase I (From Early Historic Times Until Around 1950)

During Phase I, groundwater had limited exploitation, and traditional water-harvesting methods were given impetus through unorganized efforts by local communities, who valued water and practiced its conservation, using their knowledge of the terrain, topography, and hydrogeology of their areas, to meet local water requirements at times of crisis. In those days, water harvesting was a part of the sociocultural framework. There are stone inscriptions (600–1000 CE) describing construction of ponds/tanks to collect rainwater to recharge wells to augment surface supplies and to provide drinking and irrigation water (DHAN Foundation 2002; Bhalge and Bhavsar 2007). Some of these structures survive even today, and are in use for multiple purposes in the southern coastal towns and villages of Tamil Nadu. More than 500 000 tanks/ponds exist all over the country.

4.2.2 Phase II (1950 to the Green Revolution in 1960)

During the 1950s, in the face of highly erratic monsoons, rational scientific knowledge and technological modernization were seen as panaceas and dominated decision makers' views on agricultural and irrigation development. To provide water and electricity to villages and towns, particularly in semi-arid regions, the emphasis was on construction of large-scale irrigation projects, big dams and canals, to eradicate rural "backwardness" and poverty, with centralization of natural resource management and funding support from many international donors and development agencies. During this period, governments and NGOs provided a limited scientific knowledge base for putting artificial recharging of groundwater into practice.

4.2.3 Phase III (1960–1970)

Phase III started with the Green Revolution and large-scale overextraction of groundwater, especially in arid and semi-arid regions. Both the government and public began to realize the importance of recharging aquifers to arrest groundwater decline. Scientific knowledge and modern technology were seen as the basis for agriculture and irrigation development. For better understanding of groundwater, many universities in India introduced courses on hydrogeology and groundwater engineering. Guidelines for artificial recharge were formulated. A number of central and state water agencies, water supply and drainage boards, agricultural and other academic research institutions and NGOs carried out pilot studies on artificial recharge, its technical feasibility, and recovery of recharged water.

4.2.4 Phase IV (1970–1990)

Besides the emphasis on rational scientific knowledge generation and technology modernization, the 1970s also faced questions about the rationality of science and certainty of modernization from postmodern social sciences scholars and grassroots environmental activists, who started debates on "access to" and "control over" local natural resources. "Traditional" technology, which was earlier seen as a hindrance to modernization, came to be recognized as useful. By the late 1980s, decentralization and local control of natural resources management was receiving more emphasis. International agencies provided significant funding support to NGOs for community irrigation and participatory watershed management.

4.2.5 Phase V (1990–2000)

In the 1990s, both public and government became more aware of the significance of artificial recharge, due to campaigns by NGOs and media, high incidence of droughts, declining groundwater levels, and water scarcity in many parts. Community-based natural resource management became very popular within international agency circles and their interventions were used to further legitimize the "new traditionalism" (Sinha et al. 1997). The NGOs, with support from the public, philanthropists, international agencies, and religious leaders, started sponsoring projects for artificial recharging through dug wells and bore wells, check dams and percolation ponds, and implementing participatory watershed management schemes on a large scale (Shah 1998). State governments also became enthusiastic about promulgating groundwater extraction regulation acts in metropolitan areas and made the implementation of rainwater harvesting and recharging schemes mandatory.

4.2.6 Phase VI (2000–Present)

During the past two decades, due to continued decline of groundwater levels and water scarcity in many area, failure to arrest groundwater decline, lack of maintenance or declining conditions of a number of traditional water harvesting systems (TWHS), the need to revive these factors has dominated the development discourse. The planning protocol under the Groundwater Recharge Management Plan (GRMP) investments in a basin/region follows the supply-side principle "recharge groundwater in places where the volume of 'uncommitted surplus water' is most available, and where storage space of the aquifers is the largest". But the GRMP has not paid much attention to the demand-side principle "recharge as much as feasible where aquifers are most depleted, and groundwater most intensively used, if needed, by reallocating water from existing uses that are less productive at the basin level than groundwater use". "Uncommitted surplus water" available is estimated from the total available runoff minus the requirements of large and small surface storage structures (existing, under construction, and planned in future).

However, in most parts of India, groundwater demand is booming and aquifers come under stress mostly in places where neither "uncommitted surplus water" nor roomy aquifers are available. In the context of India's groundwater economy, a logical criterion to direct resources would be the degree of groundwater overexploitation. The Central Ground Water Board (CGWB) categorizes blocks and districts into white (safe for development), gray (semi-critical), dark (critical), and overexploited (withdrawals exceeding long-term recharge) according to the development of known groundwater potential. Dark and overexploited blocks reflect a crisis situation, needing immediate ameliorating measures. Logically, more resources – financial as well as water – should be allocated to groundwater recharge in these blocks.

4.3 Methods of Rainwater Harvesting

4.3.1 Direct Methods

- Surface spreading techniques
  - a) Flooding
  - b) Ditch and furrows

    c) Recharge basins;
    d) Runoff conservation structure:
        i) Bench terracing
        ii) Contour bunds and contour trenches
        iii) Gully plugs, nala bunds, check dams
        iv) Percolation ponds
    e) Stream modification/augmentation
- Subsurface techniques
    a) Injection wells (recharge wells)
    b) Gravity head recharge wells
    c) Recharge pits and shafts

4.3.2 Indirect Methods

- Induced recharge from surface water sources
- Aquifer modification:
    i) Bore blasting
    ii) Hydro-fracturing

4.3.3 Significance of Groundwater Aquifers

Natural groundwater aquifers have great value, but they can be depleted by overexploitation if natural and artificial discharge exceeds natural recharge. The largest potential reservoir for the storage of rainwater is in the unsaturated soil zone. Use of this space for the storage and retrieval of potable water is a multifaceted problem that requires application of the best scientific approaches. To overcome groundwater depletion using surplus rainfall runoff, attempts are made to increase water availability by artificially replenishing the groundwater reservoir by installation of infrastructural facilities for water storage on land surface and through recharge wells, which may serve more than one purpose. In certain areas, for example, artificial recharge not only adds water to the available groundwater supply, but also is a means of disposing of storm water runoff. In another instance, artificial recharge to control salt water intrusion also increases the available supply of fresh water and alleviates a ground subsidence condition that has been in progress for years. Both of the above-mentioned techniques require a thorough knowledge and understanding of the land, geology, and hydrology of the area.

Compared to water storage on the land surface, groundwater storage has many advantages such as no water losses by evaporation, reduced construction cost in preparing the surface storage, and off-season availability of water, by increasing water in a depleted aquifer, usually accomplished during the season. Different methods have been developed and applied to artificially recharge groundwater in various parts of the world and in India (Todd 1980; Asano 1985; CGWB 1994; Agrawal and Narain 2001). Efforts are being made for harvesting rainwater and groundwater recharge, both in rural areas and urban areas. If such structures are not maintained properly, water quality may be a problem. In urbanized areas with very high population density, rooftop harvesting can provide very limited water. Moreover, the groundwater as well as surface runoff in such areas is highly polluted. In addition, indirect recharge enhancement may work for shallow groundwater circulation, but deeper systems require sophisticated injection and alternative sources of high-quality water (Datta 2013).

It also has to be kept in mind that only some shallow aquifers receive recharge from surface water and the deeper aquifers, which contain old water accumulated thousands of years ago, cannot be recharged from surface water sources (Datta 2013). For success of a recharge project, field conditions must provide for appropriate storage, movement, and use of recharge water, satisfying the following physical requirements for recharging.

- The basin geology must be suitable for storage capacity and transmissibility of aquifers.
- Adequate recharge water must be available.
- Recharge rates must be maintained at adequate levels.
- In places where the water table is near the surface, adequate drainage and storage capacity in the basin for recharging must be provided.
- Recharge water quality must be compatible with groundwater quality.
- Pumping lifts must not be excessive, installed pumping capacity must be efficiently used, and quality of water recovered must be satisfactory.

4.3.4 Artificial Recharge of Groundwater

Rainwater harvesting systems for artificially managing recharge to groundwater are a common approach for increasing groundwater supplies. Artificial recharge systems are established to augment groundwater resources in such places where surplus surface runoff water can be put on or in the ground for infiltration and subsequent movement to aquifers. Other objectives of artificial recharge are to store water, to improve the quality of the water through soil-aquifer treatment or geopurification, to use aquifers as water conveyance systems, and to make groundwater out of surface water where groundwater is traditionally preferred over surface water for drinking. Artificial recharge is expected to become increasingly necessary in the future as growing populations require more water, and as more storage of water is needed to save water in times of water surplus for use in times of water shortage.

There are several recharge methods, including injection wells, aquifer storage and recovery (with injection and extraction through the same wells), and infiltration basins. Injection wells and aquifer storage and recovery may offer advantages such as limited land requirements, but can be technically challenging to design, have high energy and water quality requirements, and require creation and maintenance of conveyance and pumping systems. In contrast, artificial recharge through infiltration basins, in which water is collected in natural depressions or constructed ponds/tanks for infiltration into the subsurface over time, may require less engineering and lower operating costs than injection wells and/or aquifer storage and recovery systems. The recharge projects have demonstrated improvements in water quality during the infiltration process. These improvements can be particularly important for places lacking reliable access to pristine surplus surface water supplies.

Artificial recharge efforts are basically augmentation of the natural movement of surface water into groundwater reservoirs through suitable civil construction techniques or other similar methods. Availability of source water is one of the important requirements for recharge schemes. This is assessed in terms of non-committed surplus runoff, which is going unutilized as per the water resource development pattern. The other basic requirement is the availability of subsurface storage space in different hydrogeological situations. The topography and soil condition of the area link the above two factors. Topography governs the extent of runoff and its retention whereas the soil condition determines the extent of percolation. The artificial recharge technique interrelates and integrates the source water to groundwater reservoir, which in turn is dependent on the hydrogeological situation of the area.

Artificial recharge projects are site specific. Replication of techniques from similar areas has to be based on the local hydrogeological and hydrological situation. The first step in planning is to demarcate the area of recharge. However, localized schemes can also be taken up to augment the groundwater in situations of (i) declining groundwater levels over a period, (ii) substantial desaturation of aquifer, (iii) inadequate groundwater availability in lean months and availability of surface water for recharge during rainy season, (iv) salinity ingress, and (v) quality problem in ground water.

In general, human activities that enhance groundwater recharge can be grouped in three categories.

- *Unintentional*: for example, through clearing deep-rooted vegetation, by deep seepage under irrigation areas and by leaks from water pipes and sewers.
- *Unmanaged*: for example, storm water drainage wells and sumps, and septic tank leach fields, usually for disposal of unwanted water without thought of reuse.
- *Managed*: for example, through injection wells, infiltration basins and galleries for rainwater, storm water, reclaimed water, mains water and water from other aquifers that is subsequently recovered for all types of uses.

4.3.5 Recharging Groundwater by Rooftop Rainwater Harvesting

Artificial recharging of groundwater by storage of water in underground tanks for domestic supplies is common in urban and rural areas of India, using various types of structures. Commonly used methods include (i) recharging of bore wells; (ii) recharging of dug wells; (iii) percolation tanks; (iv) recharge trenches; (v) soakaways or recharge shafts; and (vi) recharge pits (Figure 4.1).

4.3.6 Direct Recharge

This is the simplest, most widely used method for groundwater recharge, in which water is made to infiltrate from the land surface through the soil zone to the groundwater by three methods.

- *Shallow aquifer*: water spreading over fields or conveyed to basins and ditches, runoff conservation structures (bench terracing, contour bunds and contour trenches, gully plugs, nala bunds, check dams, percolation ponds), and stream modification/augmentation practiced on large scale all over the world.
- *Deep aquifer*: recharge by flooding pits and dug shafts; this has limited application, as recharge capacity is low; however, abandoned stone quarries/open wells can help, where the water table has declined below the excavated depths.
- *Confined aquifers*: recharge by injecting surface water directly into the aquifer using boreholes or tube wells; this may be capital and technology intensive.

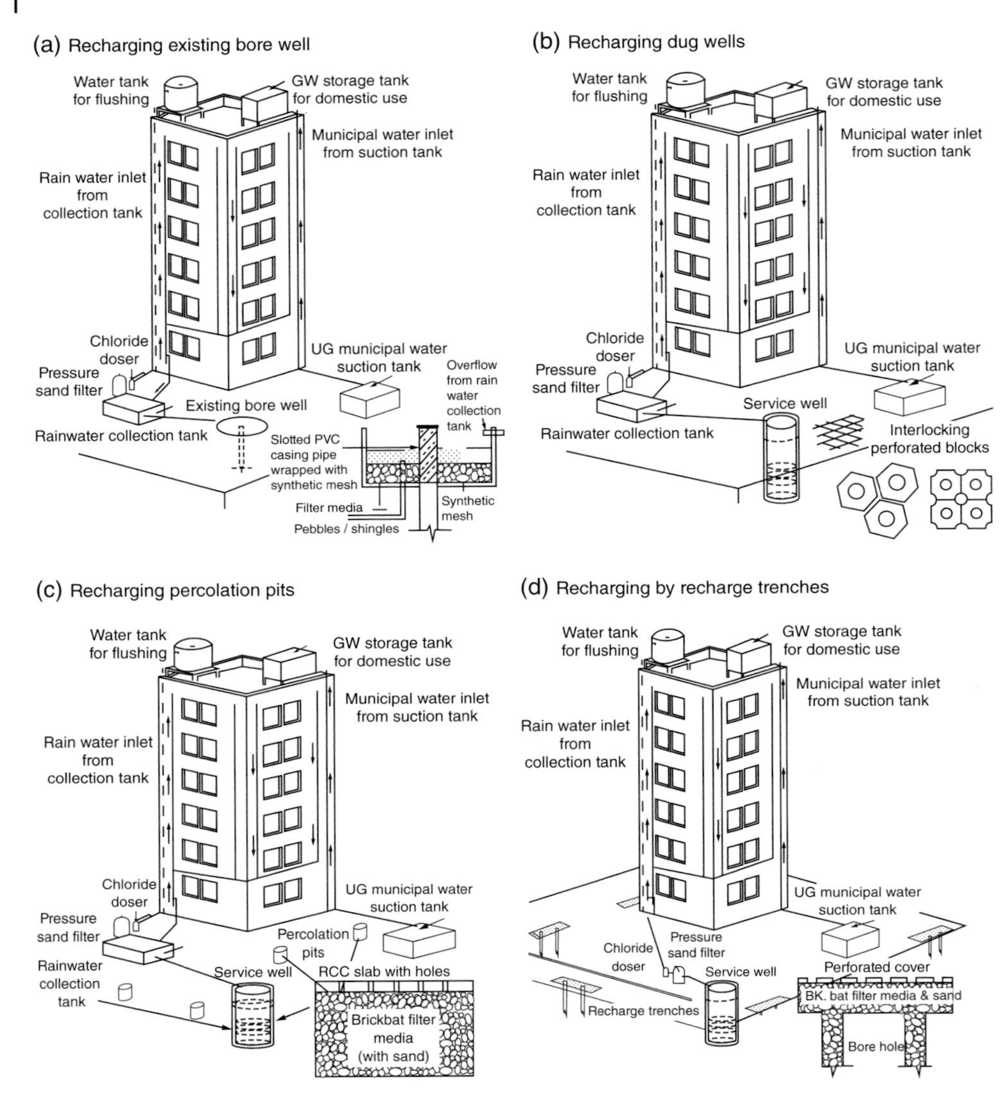

Figure 4.1 Recharging groundwater by rooftop rainwater harvesting using (a) existing bore well, (b) dug well, (c) percolation pits, and (d) recharge trenches.

4.3.7 Indirect Recharge

This is induced recharge from surface water sources (e.g. streams, canals, lakes, irrigation tanks, irrigation canals and river courses, flood irrigation, etc.) using groundwater pumping or infiltration galleries near hydraulically connected groundwater and surface water bodies to lower groundwater levels and induce infiltration from surface water bodies. Other approaches include aquifer modification (bore blasting, hydro-fracturing). In India, taking advantage of easy availability of pumping machinery at affordable prices and subsidized power, many enterprising farmers resorted to pumping induced recharge water from wells located near surface water, and transporting it long distance through pipelines to recharge their own wells to irrigate non-command areas of orchards and other high-value crops. In the Saurashtra region, an indirect recharge scheme covering 98 000 wells was created (Parthasarathi and Patel 1997).

4.3.8 Recharge by Integrated Water Resource Management

This involves increasing recharge from reservoir and canal seepage, injection and infiltration of irrigation return flow, enhanced infiltration of rainfall by leveling of irrigation fields, and basin development involving construction of check dams and minor irrigation dams.

4.3.9 Macro- and MicroCatchment Water Harvesting (MacroWH, MicroWH)

Rainwater runoff from a catchment area, such as natural slopes, shallow soils, sealed and compacted surfaces, mountains or hillsides, roads, rocky areas, rangelands, cultivated and uncultivated land, is collected into an application area where it accumulates in natural or man-made depressions, holes, pits, basins, and bunds, and/or even in ephemeral riverbeds. The runoff is conveyed through overland, rill, gully or channel flow, and either diverted or spread onto cultivated fields (where water is stored in the soil) or into specifically designed storage facilities. Soils need to be deep enough to construct holes and pits to store the collected water. Soils with an inclination to sealing and crusting are particularly suitable for inducing runoff in the catchment area. Furthermore, MacroWH can be applied on highly degraded soils where it can be used in the productive rehabilitation process, reducing erosion and flooding. Such systems, if applied over a wide area, may recharge groundwater significantly.

MicroWH is suitable for semi-arid to arid areas with high rainfall variability within seasons. These systems are designed to collect runoff from a relatively small natural catchment area (10–500 m^2), usually within the farm boundary. The ratio between the catchment (collection) area and the cultivated (application) area can vary between 2:1 and 10:1. The runoff coefficient is relatively low (0.1–0.5, i.e. 10–50% of annual rainfall); the coefficient decreases with the length of the catchment.

Most MacroWH practices have a catchment length of usually 30: 200 m, and area of <2 ha, but in some cases runoff is collected from catchments as large as 200 ha. The most common technologies are planting pits, micro-basins, triangular/V-shaped bunds, semi-circular bunds, eyebrow terraces, WH basins, and cross-slope barriers (e.g. vegetative barriers/strips, tied ridges, contour bunds ridges, stone bunds, contour bench terraces).

In places where runoff is directly diverted to fields, the application area is identical with the storage area, as plants can directly use the accumulated soil water. The application or cropping area is either terraced or located in flat terrain. The ratio of the catchment to the application area (usually cultivated) varies between 10:1 and 100:1. The amount of rainfall and the catchment size in floodwater harvesting (FloodWH) is greater than for MacroWH, and distinguishing the two methods depends on concentration/size of runoff which is tapped. FloodWH collects water from the channel flow, while MacroWH collects sheet and rill flow and short-distance channel flow. The harvested water is mainly suitable for annual and perennial crop plants, such as sorghum, millet, maize, shrubs, trees or fodder, tolerant of temporary waterlogging or rapidly maturing on residual moisture, and livestock production but also for domestic use, depending on the quantity and quality. Water stored in the soil is directly used for plant and crop growth, prolonging the growing season and bridging the dry spells, allowing production of crops without demanding irrigation systems.

MacroWH practices are applicable in arid, semi-arid to subhumid zones where it is necessary to store water to bridge the dry season or to mitigate the impact of dry spells. The most common storage facilities include hillside runoff/conduit systems; large semi-circular or trapezoidal bunds (earth or stone); road runoff systems; and a broad range of open or closed structures. Open water storage includes farm ponds and different types of dams (often earth dams), and pans. Closed structures include groundwater dams (subsurface, sand, and percolation dams) and above- or below-ground tanks (cisterns) or reservoirs, and horizontal and injection wells. Such storage structures are often characterized by multipurpose use, prioritizing domestic and livestock consumption. During short dry spells, the stored water may sometimes be used for supplementary irrigation.

4.3.10 Managed or Artificial Recharging Methods

Although artificial recharge is a potential means of solving some water supply problems, each approach must be evaluated and assessed for its physical and economic feasibility. Geological and hydrological conditions that may affect the recharge must be evaluated in each location where artificial recharge is to be used. The quality of recharge water must also be analyzed for its chemical compatibility with the groundwater and whether it requires pretreatment to avoid clogging of the aquifer. With appropriate pretreatment before recharge and sometimes post treatment on recovery of the water, it may be used for drinking water, industrial water, irrigation, toilet flushing, and sustaining ecosystems. Finally, the most suitable method of recharge for the application must be selected and its cost determined.

4.3.10.1 Surface Water–Groundwater Interactions

Common methods for recharging groundwater are by artificial surface spread of water through basins or by induced recharge from adjacent streams and lakes or through injection wells. Mathematical models on idealized conditions have dominated research in water recharge through surface basins and empirical relations, derived by experimental sequencing of recharge operations, and operational controls in the pretreatment of recharge water. Recharge by injection wells has been undertaken in a variety of hydrological environments. In Australia and Israel, efforts have been directed toward analyses of diffusion and dispersion of the injected water. Much research in the United States has covered the movement of bacteria and organic matter through an aquifer and the chemical modeling of changes in recharged water as it moves. More research is needed on the basic properties of aquifers, particularly in the unsaturated zone, and on all aspects of recharge water quality, involving interdisciplinary teams capable of integrating the geological, hydraulic, and economic aspects of the system.

Groundwater may be recharged by water from nearby surface water sources or releasing rainfall surface runoff into basins formed by excavation or by the construction of containment dikes or small dams. The horizontal dimensions of such basins vary from a few meters to several hundred meters. The most common system consists of individual basins fed by pumped water from nearby surface water sources. However, the water should be silt free to avoid the problem of sealing of basins during flooding. Even so, most basins require periodic scraping of the bottom surface when dry to preserve a percolation surface. In California, basins have been successfully built and operated in abandoned stream channels. In alluvial plains, basins may parallel existing channels with water being led into the upper basin by canal. As the first basin fills, it spills into the second. This is repeated through the entire chain of basins. From the lowest basin, excess water is returned to the main channel. By this method, spreading is accomplished on what otherwise might be considered wasteland and permits water contact over 75–80% of the gross area. Basins, because of their general feasibility and ease of maintenance, are the most favored method of artificial recharge from surface water.

Other alternatives include ditches or furrows, which are shallow, flat-bottomed, and closely spaced to obtain maximum water contact area. Gradients of major feeder ditches should be sufficient to carry suspended material through the system since deposition of fine-grained material clogs soil surface openings. Water spreading in a natural stream channel may use any of the methods described.

Whatever method of surface application is adopted, the primary purpose is to extend the time and the area over which water is recharged. In irrigated areas, sometimes irrigating with excess water during non-irrigating seasons spreads water. This method requires no additional cost for land preparation as the distribution system is already installed. Even keeping irrigation canals full will contribute to recharge by seepage from the canals. Where a large portion of the water supply is pumped, this has the advantage of raising the water table and consequently reducing power costs. Disadvantages include additional energy cost, evaporation losses, and leaching of soil nutrients.

surface recharge may be economical depending upon maintenance of a high infiltration rate. However, typical infiltration rate curves show a pronounced decrease with time. The initial decrease is attributed to dispersion and swelling of soil particles after wetting. The subsequent increase occurs as entrained soil air is eliminated by solution in passing water, while the final gradual decrease results from microbial growths clogging the soil pores. Generally, recharge rates decrease as the mean particle size of soil on a spreading area decreases. Efforts to maintain soil infiltration rates include addition of organic matter and chemicals to the soil and vegetation cover. Alternating wet and dry periods on a basin generally will furnish a greater total recharge than does continuous spreading, in spite of the fact that water is in contact with the soil for as little as one-half the total time. This occurs because soil porosity increases during vegetation growth periods. Soils must be aerated to allow vegetation growth.

Studies of small ponds suggest that infiltration rates are directly proportional to the hydraulic head and to the permeability of material surrounding the ponded water. Clogging due to artificial recharge in laboratory-simulated, unconsolidated aquifers displays two patterns. The first, resulting from recharge with turbid water containing an effective microbial inhibitor, shows clogging throughout the aquifers ranging in depth from 48 to 123 cm. The rate of clogging at different depths depends on the size distribution of the porous media. The second, resulting from recharge with non-turbid water and no effective microbial inhibitor, shows clogging only in the top few centimeters. Water spreading cannot be effective in areas where subsurface soil strata restrict the downward passage of water.

4.3.10.2 Recharge through Pits and Tubular Wells

In areas where the impervious layer is close to the ground surface, specially designed pits and wells can conduct recharging. If these pits and wells penetrate to more permeable substrata, water can percolate directly into an aquifer.

Recharge pits are generally 1–2 m wide and 2–3 m deep, for recharging shallow aquifers. After excavation, the pits are refilled with pebbles and boulders at the bottom, gravels in between and coarse sand at the top so that the silt content that comes with runoff will be deposited on top of the coarse sand layer and can easily be removed.

Percolation pits are one of the easiest and most effective means of rainwater harvesting to make the rainwater enter directly into the aquifer. These pits are generally not more than 60 × 60 × 60 cm but the size depends on the extent of expected runoff. The excavated pit is lined with a brick/stone wall with openings (weep-holes) at regular intervals, filled with pebbles or brick jelly and river sand and covered with perforated concrete slabs wherever necessary. The design procedure is the same as that of a settlement tank. If the depth of clay soil is more, recharge through percolation pits with bore is preferable. This bore can be at the center of the square pit and is filled with pebbles and the top portion with river sand and covered with a perforated concrete slab. Depending on the lithology, necessary casing may be provided in the recharge shaft to avoid clogging.

A recharge well (injection well, inverted well, diffusion well, disposal well) is a well that admits water from the surface to underground formations to directly recharge water into deep aquifers. Its flow is the reverse of a pumping well, but its construction should be the same. Recharge wells are suitable only in areas where a thick impenetrable or slowly permeable layer exists between the soil surface and the aquifer. In a modified injection well (generally a borehole of 500 mm diameter), drilled to the desired depth depending upon the geological conditions, preferably 2–3 m below the water table, water is not pumped into the aquifer but allowed to percolate through a filter bed of sand and gravel. Inside this hole, a slotted casing pipe of 200 mm diameter is inserted. The annular space between the borehole and the pipe is filled with gravel and developed with a compressor till it gives clear water. To stop the suspended solids from entering the recharge tube well, a filter mechanism is provided at the top. Wells and boreholes can tap groundwater from the same aquifer or feeding natural springs.

Well recharging is practical in many geological environments where aquifers must be recharged, and where economy of surface space, such as in urban areas, is an important consideration. If water is admitted into a well, a cone of recharge will be formed which is similar in shape but is the reverse of a cone of depression surrounding a pumping well. By comparing the discharge equations for pumping and recharge wells, it might be anticipated that the recharge capacity would be equal to the pumping capacity of a well if the dimension of the recharge cone is equivalent to the cone of depression. However, in the field this rarely happens. The recharge rates rarely equal the pumping rates, because of the fact that pumping and recharge differ not only by a simple change of flow direction, but also by many other changes.

Although a properly designed recharge well will recharge as much as the pumping capacity, recharge water quality and turbidity may be different. Any silt carried into a recharge well is filtered out and tends to clog the aquifer surrounding the well. Similarly, recharge water may contain bacteria that can form growths on the well screen and surrounding formation, thereby reducing the effective flow area. The chemical constituents of the recharge water may differ sufficiently from the normal groundwater to cause undesirable chemical reactions, such as ion exchange in aquifers containing sizeable fractions of silt and clay. These factors all act to reduce recharge rates and, as a result, well recharging has been limited to areas where local conditions and experience have shown the practicality of the method. Generally, recharge wells cost more to construct and may recharge smaller volumes of water than do spreading areas, but may be the only practical recharge method in some circumstances. This method should be applied only when good-quality water can be used for recharge

4.3.10.3 Ponds for Groundwater Recharge

In some places, man-made depressions fill with runoff water and eventually feed underground fresh-water "lenses" floating on top of the saline aquifer (e.g. *tajamares* in Uruguay and Paraguay, *chirle* in Turkmenistan). Water pumps are used to pump the water back up to the surface. The water is used for livestock consumption and domestic use after filtration and/or chlorination but also serves to artificially recharge groundwater aquifers. Artificial recharge through infiltration ponds can be applied almost anywhere, provided that there is a supply of clean fresh water available at least part of the year, the bottom of the pond is permeable, and the aquifer to be recharged is at or near the surface.

In Bangladesh, fresh water bubbles (in brackish groundwater) are created through infiltration wells by infiltrating pond water and rainwater below the clay layer into the shallow aquifer. This fresh water is used for domestic purposes during the dry season. In South Africa, dune infiltration basins are either excavated or formed through dams retaining the water until it has infiltrated through the basin floor, and used to enhance natural groundwater recharge for drinking water supply and protection of fresh groundwater reserves against intrusion of saline water. In Niger, oasis vegetable garden irrigation was established by lifting the groundwater table using a low ridge dam (barrage) and an infiltration basin. Low floods in the *kori* Tamgak (Iférouane) are diverted to infiltration basins.

4.3.10.4 Surface Dams

4.3.10.4.1 Small Earth and Stone Dams

Rainwater storage systems such as small dams in Ethiopia and Tanzania (known as *ndivas*) are communally constructed around foot-slopes of hills or along irrigation canals to store the runoff from ephemeral or perennial rivers. *Ndivas* are suitable for areas with 300–600 mm of rainfall. The reservoirs have neither plastered walls nor covered surfaces. The water is mostly used communally for livestock consumption, and for supplementary irrigation.

4.3.10.4.2 Check Dams

A raised wall is constructed using stone, concrete, and gabion across a gully to pond/store the stream flows behind it for irrigation purposes (using either gravity or lifting mechanism), reducing the runoff velocity and enhancing gully rehabilitation. The width of the dam wall ranges between 1 and 2 m while the height varies between 2 and 4 m, depending upon the gully depth. The length of the check dam depends on the gully width while the spacing between adjacent check dams is determined based on the availability of water and potential land area to be irrigated (Figure 4.2).

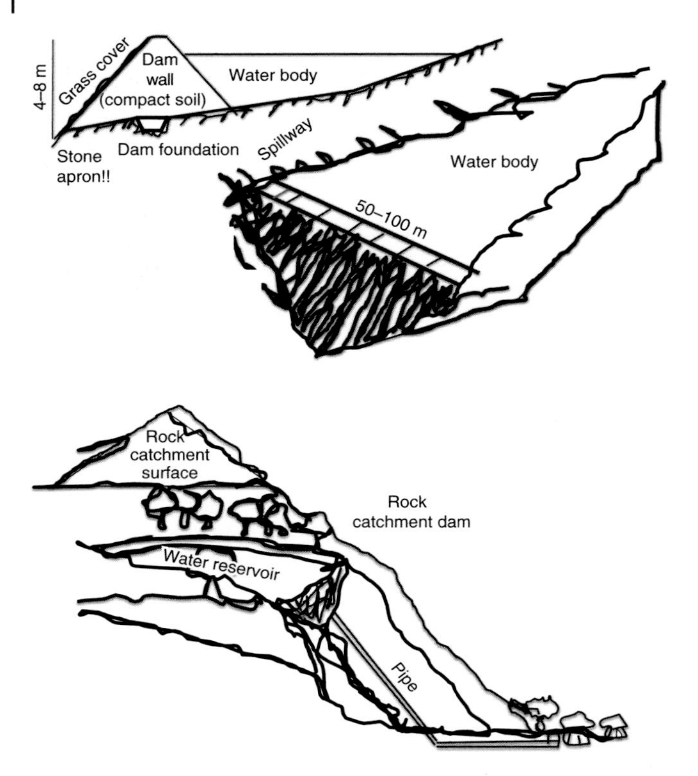

Figure 4.2 Schematics of check dam and rock catchment masonry dam.

4.3.10.4.3 Rock Catchment Masonry Dams

These dams are common practice in several countries in sub-Saharan Africa. In the case of large rock catchments, cement and stone gutters are used to extend the catchment area to gather runoff from a several-hectare catchment area. The storage structure is either a dam or a tank situated adjacent to a rock catchment. The reservoir should have a relatively high depth to surface ratio to minimize evaporation.

4.3.10.4.4 Subsurface Storage Groundwater Dams or Retention Weirs

Groundwater dams are constructed on riverbeds of sands and gravel with rock or an impermeable layer at a few meters depth, to store water below ground level and replenish the upstream dam wells. Preferably, the dam should be built where rainwater from a large catchment area flows through a narrow passage. Water is extracted for use either manually from wells or with motorized pumps. Such underground reservoirs can be filled by a single flash flood. Once saturated, the remaining water passes over the dam and replenishes downstream aquifers. The different types of dams are described below.

Percolation dams are constructed across ephemeral streams/river beds, natural drainage channels, and gullies. Simple check dams are made of locally available natural materials such as rocks, logs, bamboo, sticks, and branches. More sophisticated dams are constructed using rocks and steel rods (gabion). Concrete is used for making permanent check dams but the foundation of the dam wall must reach the impermeable layer. Crops are irrigated by pumping water from recharged wells. Subsurface dams, sand dams, and percolation dams can be combined. Raising the dam wall above the surface, causing additional accumulation of sediments, can increase the storage volume.

Sand dams are larger than subsurface dams and weirs, raised to several meters above ground in sandy ephemeral riverbeds in low-permeability lithology arid areas. Coarse sand carried by the flood flow is deposited upstream of the dam and gradually fills the streambed while lighter material is carried over during high flow. Water is stored within the porous space of the deposited coarse sand, creating an "aquifer" which can be tapped by wells in dry seasons. The thickness of this artificial aquifer increases over time when water flow occurs, and following successive floods the sand dam is raised (Figure 4.3).

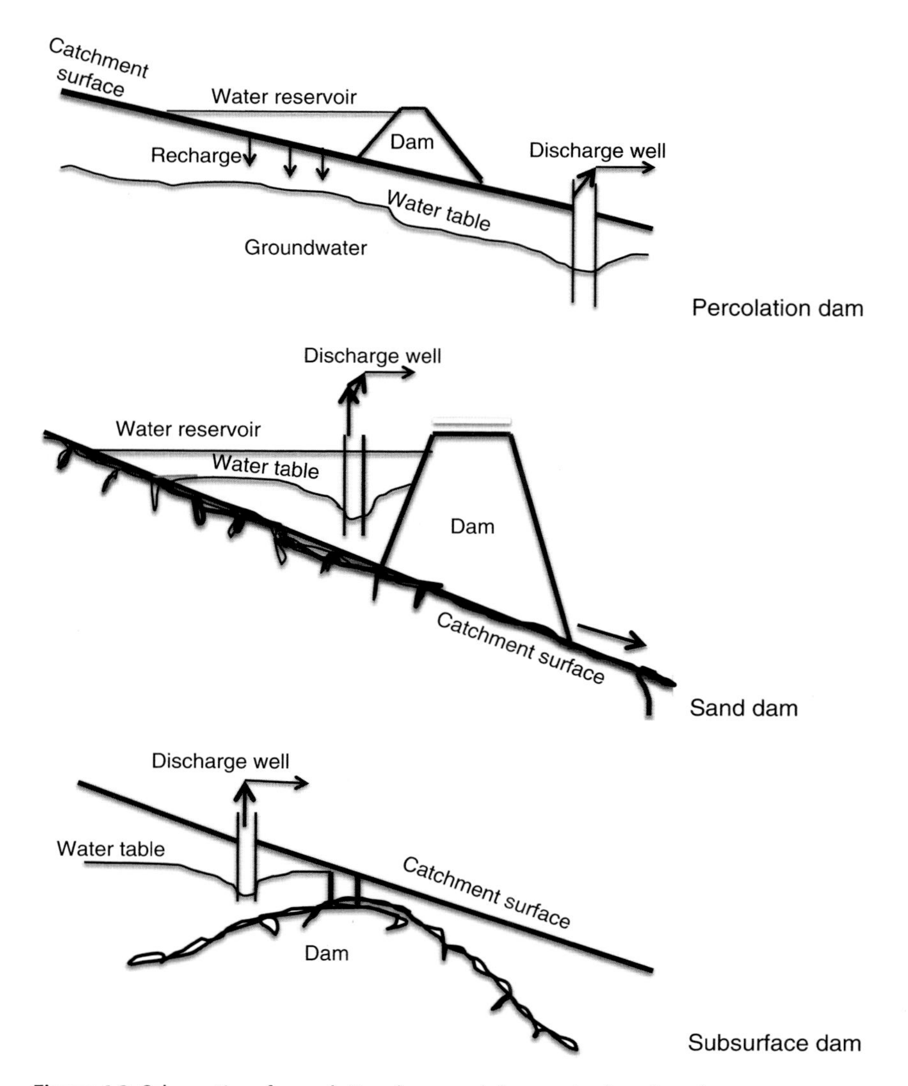

Figure 4.3 Schematics of percolation dam, sand dam, and subsurface dam.

Subsurface dams are built entirely underground into sandy riverbeds of seasonal watercourses and founded on impermeable bedrock to intercept flow of groundwater. The impermeable barriers (clay, masonry or concrete) can also obstruct subsurface flow, and groundwater can be withdrawn through wells, boreholes or collector drains. Typical small dams have a storage capacity of about 10 000 m^3 (average 4 m depth, 50 m width, 500 m length). Larger dams may have 5–10 m depth and width of 200–500 m or more, and can store 100 000–1 000 000 m^3. Several dams built in a cascade increase the total groundwater volume stored and limit the effects of leakage. Subsurface dams reduce variation in the level of the groundwater table upstream of the dam.

4.3.10.4.5 Subsurface Reservoirs or Cisterns

Cisterns are subsurface water reservoirs or storage tanks with capacity of 10–1000 m^3 dug below a solid rock layer. Large community cisterns, lined with compacted earth, clay, mortar coating, or concrete/plastic sheets to avoid seepage, can store up to 80 000 m^3 of water. When larger storage volumes are needed, two or more structures can be built in the same location. Runoff is collected from an adjacent catchment or channeled from a distant catchment. They are covered to reduce evaporation. In most cases, sedimentation traps are attached in front of the inlet to reduce clogging, otherwise regular cleaning of the cistern is required. Concrete- or ferrocement-lined subsurface tanks are known as *berkas* in Somaliland. In Turkmenistan, underground tanks built of lime mortar and bricks with a covering dome are called *sardobs.* In Gansu, China, they are called "water cellars" and in Morocco *matfia* or *joub.*

4.3.10.4.6 Horizontal Wells

Originating in Iran, these water harvesting systems created with 2500-year-old technology consist of gently sloping subterranean tunnels dug far enough into alluvium or water-bearing sedimentary rock to pierce the underground water table and penetrate the aquifer beneath. Water from the aquifer filters into the upper reaches of these channels, flows down their gentle slopes, and emerges as a surface stream of water at or near a settlement. These wells are generally constructed on the slopes of foot zone alluvial fans, in potentially fertile arid areas with high evaporation rates, close to precipitation-rich intermountain basins, and where underground springs are common, along alluvial valleys that lack large rivers with year-round flows sufficient to support households and irrigation. Water channeled to villages or farmland using gravity saves labor compared to obtaining water from dug wells using manual labor. The rehabilitation and maintenance of these systems require great knowledge and skills. The technology goes by different names: *faladsch/aflaj* (United Arab Emirates and Oman), *foggara* (North Africa), *galerias* (Spain), *kanjering* (China), *karez* (Afghanistan, Paksitan), and *qanats* (Syria, Jordan).

4.3.10.4.7 Managed Aquifer Recharge Methods

Managed aquifer recharge (MAR) methods currently in use in Australia include the following.

- *Aquifer storage and recovery (ASR)*: useful in brackish aquifers, this technique involves injection of water into a recharge well for storage in groundwater and recovery from the same well. During the recharging process, upwelling of the water table occurs, and during recovery a cone of depression is created in the water table.

- *Aquifer storage, transfer, and recovery (ASTR)*: this involves injecting water into a recharge well for storage, and water recovery from a different well, to achieve additional water treatment in the aquifer by extending residence time in the aquifer. During the recharging process, upwelling of the water table occurs in the recharging well, and during recovery a cone of depression in the water table is created in the recovery well.
- *Infiltration ponds*: these divert surface water in off-stream basins and channels that allow water to soak through an unsaturated zone to the unconfined aquifer. Recovery of water is done from a well adjacent to the infiltration pond.
- *Infiltration galleries*: water is collected in buried infiltration trenches (having polythene cells or slotted pipes) in permeable soils which allow infiltration through the unsaturated zone to an unconfined aquifer. Recovery of water is done from a well at a distance adjacent to the infiltration trench.
- *Soil aquifer treatment*: this involves infiltrating treated sewage effluent intermittently through infiltration ponds to facilitate nutrient and pathogen removal via passage through the unsaturated zone for recovery from the unconfined aquifer by wells adjacent to the infiltration ponds.
- *Rainwater harvesting for aquifer storage*: this involves diverting roof runoff into a well, sump or caisson filled with sand or gravel which is then allowed to percolate to the water table where it is collected by pumping from a well.
- *Recharge releases*: these involve dams on ephemeral streams to detain floodwater and slowly release water into the streambed downstream to match the capacity for infiltration into underlying aquifers.

4.3.10.4.8 Components of Managed Aquifer Recharge

The seven components that are common to all types of MAR projects are capturing zone, pretreatment, recharge, recovery, posttreatment, and end-use. However, MAR projects may be quite different for confined and unconfined main aquifers. In the case of a confined aquifer, captured water after pretreatment is injected via a recharge well, and water discharge is done via the same well. During water recharge, upconing of piezometric level occurs, and during discharge a cone of depression of piezometric level occurs. In the case of an unconfined aquifer, captured water after pretreatment is made to infiltrate through a basin through permeable soils to the aquifer, where recharge can be enhanced by basins and galleries. Water recharging through the infiltration basin causes a water table mound and storage beneath. In the case of both types of aquifers, the subsequent steps of recovery, posttreatment and end-use are the same, and during water recovery from discharge wells, groundwater lateral flow takes place.

There are many combinations of water sources, water treatments, and end uses. Generally, poorer quality source waters will need a higher level of treatment before recharge in cases where (i) the aquifer already contains high-quality water, (ii) the water is to be recovered for higher value uses such as drinking, or (iii) the aquifer is fine-grained and there is a need to avoid frequent or permanent clogging of the recharge basin, gallery or well.

Other forms of MAR that were/are used in other countries include the following.

- *Dry wells*: these are shallow wells where water tables are very deep, allowing infiltration of very high-quality water to the unconfined aquifer at depth (e.g. Phoenix, USA).

- *Bank filtration*: extraction of groundwater from a well or caisson near or under a river or lake to induce infiltration from the surface water body, thereby improving the recovered water quality and making it more consistent (e.g. Berlin, Germany).
- *Dune filtration*: infiltration of water from ponds constructed in dunes and extraction from wells or ponds at lower elevation for water quality improvement and to balance supply and demand (e.g. Amsterdam, The Netherlands) (Figure 4.4).

Selection of suitable sites for MAR and choice of method depend on the hydrogeology, topography, hydrology, and land use of the area. MAR can play a role in increasing storage capacity to help city water supplies cope with runoff variability in catchments exacerbated by climate. It can also assist in harvesting abundant water in urban areas that is currently unused.

Common reasons for using MAR include:

- securing and enhancing water supplies
- improving groundwater quality

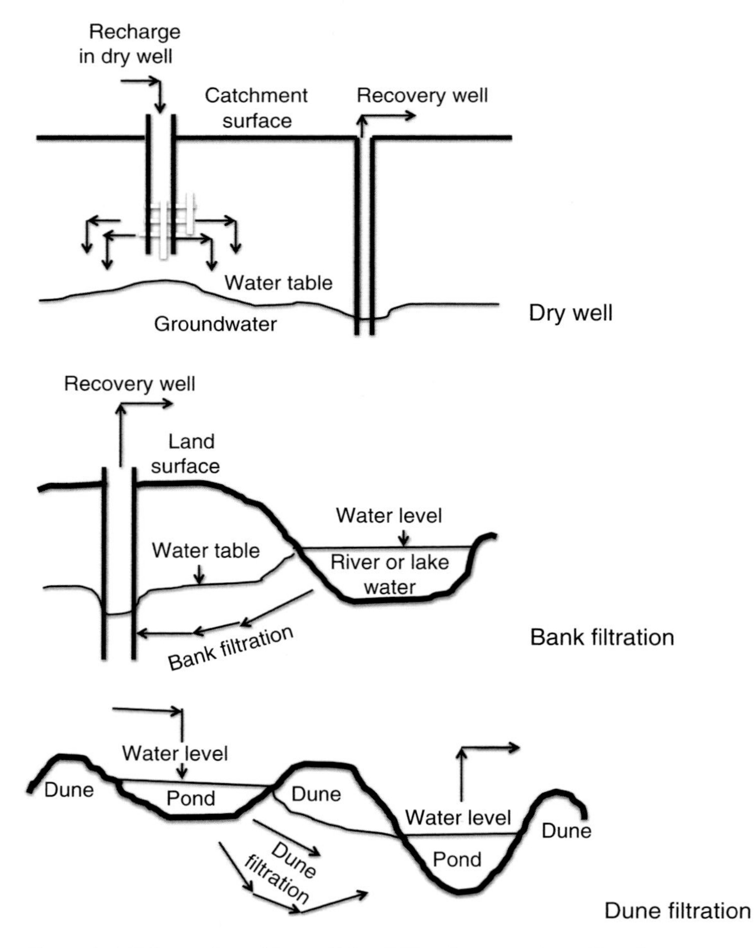

Figure 4.4 Schematics of dry well, bank filtration, and dune filtration.

- preventing salt water from intruding into coastal aquifers
- reducing evaporation of stored water
- maintaining environmental flows and groundwater-dependent ecosystems, which improve local amenities, land value, and biodiversity.

Consequential benefits may include:

- improving coastal water quality by reducing urban discharges
- mitigating floods and flood damage, or facilitating improvements in urban landscape that increase land value.

4.3.11 Worldwide Use of Water Harvesting Structures

4.3.11.1 Groundwater Dams

- *Subsurface dams*: East Africa (e.g. Ethiopia, Kenya, Tanzania), Australia.
- *Sand dams*: found throughout the semi-arid regions; highest concentration in Kenya, also found in Angola, South Africa, Sudan, Uganda, Zimbabwe, Namibia, Australia, Japan, India, Thailand, south-west USA, and Brazil.
- *Percolation dams*: widely used in Saudi Arabia, United Arab Emirates and Oman, Egypt, India, Jordan, Peru, Sudan, Syria, Thailand, Australia, and Yemen.
- *Gully plugging/productive gullies*: Bolivia, Ethiopia, Haiti (*jardin ravines*), India, Kenya, Morocco, Nepal, Nicaragua, Tajikistan, and Tanzania.

4.3.11.2 Ponds

- *Hafirs:* savannah belt of Africa (e.g. Ethiopia, Kenya, Morocco), Sudan, Middle East (rural Bedouin communities, e.g. Jordan).
- *Cultivation reservoirs/teras*: India and Sri Lanka, Pakistan, Somalia, Sudan.
- *Infiltration ponds*: Australia, Bangladesh, Niger, Paraguay, Uruguay, South Africa.

4.3.11.3 Surface Dams

East Africa (e.g. Burundi, D.R. Congo, Ethiopia, Kenya, Somalia, Sudan, Tanzania, Uganda, Zambia), southern Africa (e.g. Botswana), West Africa (e.g. Burkina Faso, Senegal), Latin America (e.g. Brazil, Paraguay, Peru), Asia (e.g. China, India), Israel.

4.3.11.4 Cisterns

North Africa (e.g. Egypt, Libya, Morocco, Tunisia), Middle East (e.g. Jordan, Syria, Yemen), eastern and southern Africa (e.g. Botswana, Ethiopia), Asia (e.g. India), Latin America (e.g. Brazil).

4.3.11.5 Horizontal Wells

Asia (e.g. Afghanistan, China, India, Iran, Iraq, Jordan, Pakistan, Syria), Arabian Peninsula (e.g. Oman, United Arab Emirates), North Africa (e.g. Algeria, Egypt, Libya, Morocco, Tunisia), Europe (e.g. Spain – Canary archipelago, Greece, Italy).

4.3.11.6 Recharge Wells

North Africa (e.g. Tunisia), East Africa, India, etc.

4.3.11.7 MAR Methods

Australia.

Other forms of MAR that were/are used include dry wells (e.g. Phoenix, USA), bank filtration (e.g. Australia; Berlin, Germany), and dune filtration (e.g. Australia; Amsterdam, The Netherlands).

4.3.12 Traditional Water Harvesting Systems in India

Various types of RHWS exist, including kunds, kuis, khadin, nadis, tanaks, talabs, beri, virdas, johads, baoli, etc. in the Thar Desert and Gujarat; bamboo pipes and apatani systems in the eastern Himalayas, ghul in the western Himalayas; zabo and cheo-ozihi in north-eastern India; dongs, garh, and dara in the Brahmaputra Valley; the havelis of Jabalpur; bandh and bandhulia in Satna; ahar-pynes in Bihar; eri and kulam in eastern coastal plains; and jackwells on the islands.

4.3.12.1 Kunds/Kundis

These are water harvesting pits with a dome-shaped cover, made of local pond silt, charcoal ash and small gravels or cement, having a saucer-shaped catchment area gently sloping to the center where the well is situated. A wire mesh across the water inlet prevents debris falling into the pit. Kunds are more prevalent in western arid Rajasthan, where limited groundwater of moderate to high salinity is available, and mostly owned by private households or caste groups (Dhiman and Gupta 2011).

4.3.12.2 Kuis/Beris

Kuis are very narrow-mouthed, 10–12 m deep pits dug in the vicinity of tanks to collect the seepage, to harvest rainwater in areas with meager rainfall in western Rajasthan. The openings of the structures are generally covered with planks of wood, or put under lock and key (Dhiman and Gupta 2011).

4.3.12.3 Nadis

These are village ponds for storing rainwater runoff from adjoining natural catchment areas, found near Jodhpur in Rajasthan, having storage capacity to retain water for 4–8 months in normal rainfall years, and even throughout the year, depending on the catchment and runoff characteristics (CGWB 2011).

4.3.12.4 Stepwells or Baoli

These usually consist of covered and protected wells/ponds with a vertical shaft from which water is drawn, with surrounding inclined subterranean passageways, chambers and descending steps on three/four sides which provide access to the well, often rectangular in design and of architectural significance. They may be multi-storied, using bullock and water wheels to raise the water in the well to the first/second floor. The stepwells originated from the need to ensure water during the period of drought when water level declines. Most of the surviving covered stepwells also serve for leisure purpose, as well as providing water (Shekhawat 2015) and relief from heat, as a place for social gatherings and religious ceremonies. These are most common in Gujarat and Rajasthan and are given the local name *"jhalara"* (human-made tanks), essentially meant for community use and religious rites. Jhodhpur has eight jhalaras, the oldest being Mahamandir (1660 CE).

4.3.12.5 Virdas

These are shallow holes, made in the sands of dry riverbeds and lakes, for collecting drinking water. They are found all over the Banni grasslands of the great Rann of Kutch in Gujarat. The sweet fresh water remains in the upper layer and the saline water stays below the fresh water. The Maldharis (local nomadic people) first established these structures in the Rann of Kutch (Centre for Science and Environment 2011).

4.3.12.6 Tankas/Tanks

These are underground circular holes in the ground used to collect rainwater, with large walls on four sides, lined with fine polished lime and with an almost impermeable floor. They have a large catchment area and a system of canals, with enormous water-holding capacity. The tanks are mostly built in houses or courtyards of Bikaner, and are often beautifully decorated with tiles to keep the water cool. The oldest tank in Jodhpur is Fatehsagar (1780 CE) (Dhiman and Gupta 2011).

4.3.12.7 Talabs/Ponds

These are water-harvesting reservoirs situated in natural depressions and valleys. Some talabs have wells in their beds, which are called beris. As of 2001, many of the talabs were sources of potable water feeding a large number of wells and baoris (Mishra 2001). The oldest talab in Rajasthan is Ranisar (1490 CE).

4.3.12.8 Johads

Johads have high embankments on three sides (Figure 4.5), made of simple stone and mud barriers built across the contour of the slope with the fourth side left open to store rainwater for multiple uses and to recharge groundwater. A total of 8600 johads have been built in 1086 villages of the Alwar district in Rajasthan. Reportedly, there are over 200 johads in the catchment of the Aravari River, meeting the water needs of many villages in the state. Johads are cheaper compared to "anicuts" (see Figure 4.5), which have strongly built stone/concrete barriers. Many anicuts have modern structures such as dams.

4.3.12.9 Sunken Streambed Structure (Doh)

Dohs are rectangular excavations in seasonal streambeds (Figure 4.6), generally built in sequence a few meters apart. Dohs are built in semi-arid areas with low and seasonal rainfall. They are used in conjunction with shallow wells to provide temporary storage of runoff, for increasing water yields from nearby shallow wells for supplementary irrigation of annual crops in fields adjoining the stream. The dimension of a typical doh is 1.0–1.5 m deep with variable length (up to 40 m) and width (up to 10 m), depending on streambed section, with average capacity of 400 m^3. The excavated material is deposited along the stream banks as a barrier against siltation from surrounding areas. The slopes of the excavation are gentle (an upstream slope of 1:6 or 17% and a downstream slope of 1:8 or 12%) so that water flows into it and excess water out again, carrying silt rather than depositing it. The sides are steep, to increase capacity, and would benefit from stone pitching to stabilize them. A silt trap comprising a line of loose boulders is constructed across the streambed upstream.

As a supportive measure, the catchment area is treated with gully plugs (small stone checks in gullies). A water-harvesting tank (small reservoir or dam) may be excavated

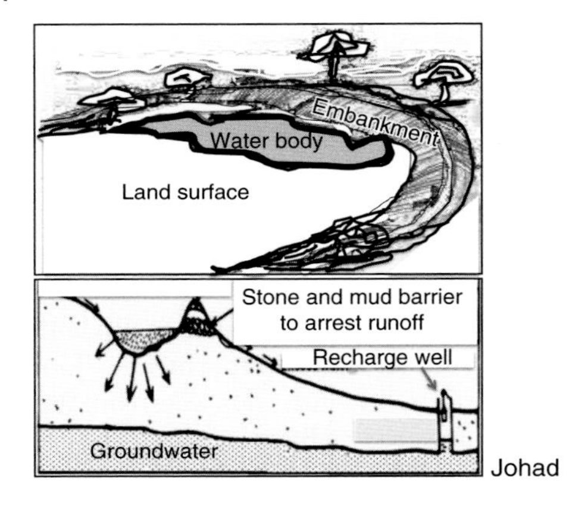

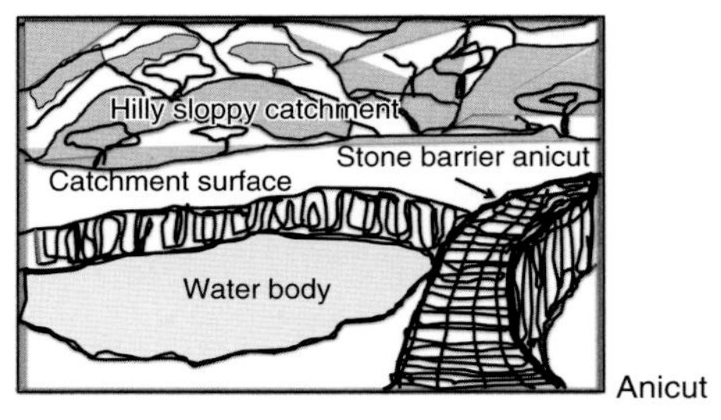

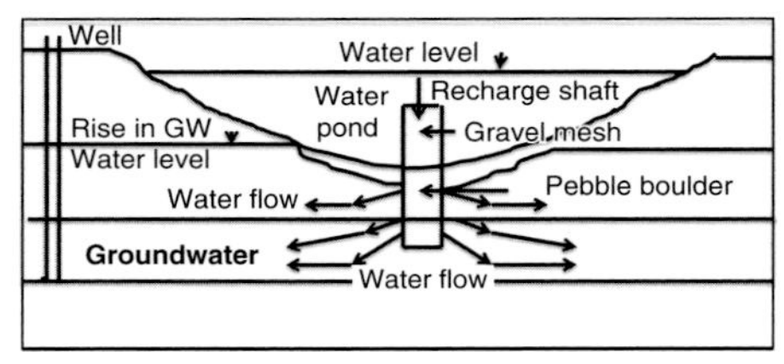

Figure 4.5 Different traditional water harvesting structures in India.

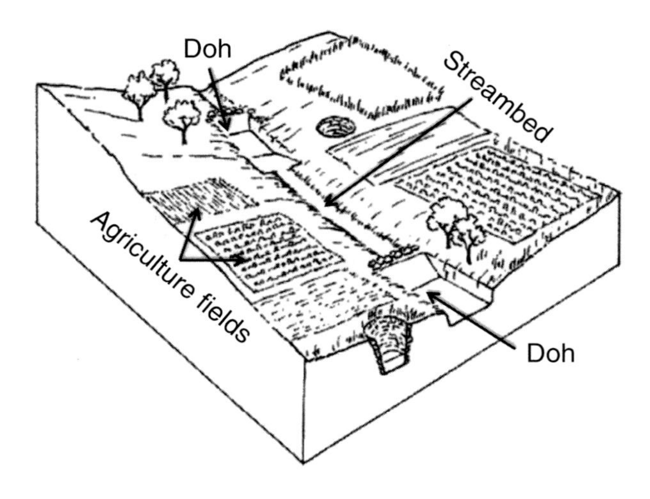

Figure 4.6 Schematic representation of sunken streambed structures (dohs).

above the series of dohs where this is justified by a sufficiently large catchment area/suitable site. The construction cost of one doh depends on its size (about 1 cubic meter can be excavated per person per day). On a per hectare basis, the costs are very variable, since they are related to the extra area brought under irrigation. Where there is underlying rock, mechanical drills and blasting by dynamite may be required, which increases the costs. The cost of deepening/widening the wells can be reduced by involving local villagers.

4.3.13 Recharge Wells

Recharge wells began to be used especially during World War II. Later, their use was extended to sea intrusion control, treated waste water, water harvesting in dry areas, and storage to recharge deep groundwater aquifers. Recharge wells are used in combination with gabion check dams to retain water to make it flow through the recharge well and to enhance the infiltration and percolation of floodwater into the aquifer. In areas where the permeability of the underlying bedrock in front of a gabion is too low, recharge wells could be installed in wadi (ephemeral river) beds.

A recharge well comprises a drilled hole, up to 30–40 m deep, that reaches the water table, a long inner tube surrounded by an outer tube (circumference 1–2 m) and a surrounding filter to allow direct injection of floodwater into the aquifer. The area between the tubes is filled with riverbed gravel as a sediment filter. Water enters the well through rectangular-shaped openings (~20 cm long, a few mm wide) located in the outer tube, and flows in the inner hole having passed through the gravel and the rectangular openings of the drill hole. The aboveground well height is around 2–3 m.

The drill hole is connected either directly with the water table or indirectly via cracks. Pond volume depends on the gabion check dam size, generally 500–3000 m^3. The floodwater retained behind the gabion check dam flows through the outer tube and the gravel filter into the water table. In order to determine potential sites, a hydrogeologist and a soil and water conservation specialist should be involved. Drilling needs to be carried out by a specialized company.

4.3.14 Coastal Reservoirs

This RWH method of building reservoirs at river mouths aims to collect only floodwater from coastal catchment, when the water is relatively clean and the waste-water content is very low, but not suitable for groundwater recharge. The key challenge for the collection of river runoff is keeping river water from mixing with sea water, as saline water tends to mix with fresh water in estuaries under the influence of waves and tides. Solid barriers or embankments separate fresh water from sea water, so that the fresh water can be contained and stored in the sea.

Generally, coastal reservoirs are vulnerable to contamination as coastal areas are rich in pollutants, and the coastal reservoirs that use solid dams tend to fail to provide good-quality water. Using tidal gates and gates to the coastal reservoirs can possibly solve the water pollution problem. If the incoming water is heavily polluted, the river water can be discharged into the sea as the tidal gates will be opened and the gates to the coastal reservoirs will be closed. If the incoming water is very clean (flooding periods), the tidal gates can be closed, keeping the gates toward the coastal reservoirs open, so the clean water can be stored in the reservoirs.

References

Agrawal, A. and Narain, S. (2001). *Dying Wisdom*. New Delhi: Centre for Science and Environment.

Asano, T. (1985). *Artificial Recharge of Groundwater*. Boston: Butterworth Publishers.

Bhalge, P. and Bhavsar, C. 2007. Water Management in Arid and Semi Arid Zone: Traditional Wisdom. International History Seminar on Irrigation and Drainage, Tehran, Iran, pp. 423–428.

Central Ground Water Board (CGWB) (1994). *Manual on Artificial Recharge of Groundwater*, Technical Series: Monograph No. 3. Government of India, New Delhi: Ministry of Water Resources.

Central Ground Water Board (CGWB) (2011). *Select Case Studies Rain Harvesting and Artificial Recharge*, 10–11. Ministry of Water Resources, New Delhi: Central Groundwater Board.

Centre for Science and Environment (2011). *Report*. New Delhi: Centre for Science and Environment.

Datta, P.S. (2013). Ensemble models on palaeoclimate to predict India's groundwater challenge. *Acque Sotterranee – Italian Journal of Groundwater* 2 (3): AS06027: 007–018.

DHAN Foundation. (2002) Revisiting Tanks in India. National Seminar on Conservation and Development of Tanks, New Delhi.

Dhiman, S.C. and Gupta, S. (2011). *Rainwater Harvesting and Artificial Recharge*. New Delhi: Central Ground Water Board, Ministry of Water Resources.

Mishra, A. (2001). *The Radiant Raindrops of Rajasthan*. Dehradun: Research Foundation for Science, Technology and Ecology.

Parthasarathi, G.S. and Patel, A.S. (1997). *Groundwater Recharge Through People's Participation in Jamnagar Region*, 51–56. Nagpur: Indian Water Works Association.

Shah, T. (1998). *The Deepening Divide: Diverse Responses to the Challenge of Ground Water Depletion in Gujarat*. Anand: Policy School.

Shekhawat, A. (2015) Stepwells of Gujarat. Available at: www.indiasinvitation.com/stepwells_of_gujarat.

Sinha, S., Gururani, S., and Greenberg, B. (1997). The 'new traditionalist' discourse of Indian environmentalism. *Journal of Peasant Studies* 24 (3): 65–99.

Todd, D.K. (1980). *Groundwater Hydrology*, 2e. New York: Wiley.

Further Reading

Borthakur, S. (2008). Traditional rainwater harvesting techniques and its applicability. *Indian Journal of Traditional Knowledge* 8 (4): 525–530.

Chawla, A.S. (2000). *Ground Water Recharge Studies in Madhya Ganga Canal Project*. Colombo: Consultancy Report to IWMI.

Datta P.S. (2014). Need for Better Groundwater Governance for Sustained Water Supply. IPHE National Conference on Piped Water Supply and Sewerage Systems, January 24–25, New Delhi.

Datta P.S. (2015). Traditional Rainwater Harvesting at Crossroads – Time to Seek Ethical Opportunities for Reform. Proceedings of the Aqua Foundation IX World Aqua International Congress, Reviving Traditional Water & Environment Conservation Techniques, November 26–27, New Delhi, pp. 93–115.

Dinesh, K.M., Shantanu, G., Ankit, P. et al. (2006). Rainwater harvesting in India: some critical issues for basin planning and research. *Land Use and Water Resources Research* 6: 1–17.

Dillon, P., Pavelic, P., Page, D. et al. (2009). *Managed Aquifer Recharge*. CSIRO Australia: National Water Commission.

Eroksuz, E. and Rahman, A. (2010). Rainwater tanks in multi-unit buildings: a case study for three Australian cities. *Resources, Conservation and Recycling* 54 (12): 1449–1452.

Huisman, L. and Olsthoorn, T.N. (1983). *Artificial Groundwater Recharge*. Marshfield: Pitman Publishing.

Khastagir, A. and Jayasuriya, N. (2009). Optimal sizing of rain water tanks for domestic water conservation. *Journal of Hydrology* 381: 181–188.

NGWA (2014). *Managed Aquifer Recharge: A Water Supply Management Tool*. National Groundwater Association Available at: www.ngwa.org.

Raju, K.C.B. (1998). Importance of recharging depleted aquifers: state of the art of artificial recharge in India. *Journal of the Geological Society of India* 51: 429–454.

Russo, T.A., Fisher, A.T., and Lockwood, B.S. (2014). Assessment of managed aquifer recharge site suitability using a GIS and modeling. *Groundwater* 53 (3): 389–400.

Smithies, C. (2012). *Applications and Methods for Rainwater Harvesting in the Arid Zone*. Alice Springs: Centre for Appropriate Technology.

5

Case Studies of Implementing Water Harvesting

CHAPTER MENU

5.1 International Scenario

In recent years, rainwater harvesting (RWH) has attracted increased attention due to water scarcity, and RWH is used for storm water management, small-scale agricultural needs, and as an alternative supplementary (generally non-potable) water source in urban and rural areas in a wide range of climatic and socioeconomic environments. Artificial recharge facilities are operational in many countries in Europe, the United States, India, Australia, the Maghreb, the Middle East, Saudi Arabia, the United Arab Emirates, Oman, and Africa. Many of them are located near a well field, in order to ensure or increase its production, but others are primarily intended to increase the resources of an entire regional or local aquifer system. In arid or semi-arid regions, the main purpose of measures to control the high rainfall-induced irregular and ephemeral surface water flows by dams or dikes may be to increase the recharge of aquifers that already depend on this type of flow under natural conditions.

Rainwater harvesting potential is expressed as:

Catchment area (m^2) × Rainfall (mm) × Collection efficiency.

Collection efficiency is low in bare ground (0.1–0.2) and green areas (0.05–0.1). The design of RWH systems varies according to the goals for performance, although the physical components of the system (collection area, conveyance, and storage) remain largely consistent. In most cases, the most important and difficult decision is how much storage capacity to build. Although RWH sizing varies considerably in approach, and studies have investigated the effect of various design parameters (storage volume, collection area, demand, etc.) on system performance (reliability, water saving efficiency, overflow volume, etc.), limited effort has been made to determine the effect of rainfall characteristics on overall RWH system performance.

Since rainfall pattern has a strong effect on the overall functioning and performance of a RWH system but cannot be changed, designers focus on the controllable system

Water Harvesting for Groundwater Management: Issues, Perspectives, Scope, and Challenges,
First Edition. Partha Sarathi Datta.

components and parameters, such as the collection area, storage volume, and demand level. For cost-effective RWH solutions to water scarcity, particularly in developing countries, economic analysis of the RWH system plays an important role. However, limited scientific literature is available on economic analysis of RWH systems, often with conflicting results.

There have been many studies highlighting the benefits of harvesting rainwater in arid regions. Most of the studies reported are RWH from roof-yard catchment systems, claiming that the quality of harvested rainwater mostly met WHO standards for drinking water in terms of measured chemical constituents, even for rainwater collected immediately after rainfall. However, the fecal coliform count exceeded the limits for drinking water. Many evaluations on a number of water sources, including RWH, water storage tanks, subsurface and various surface reservoirs, suggested RWH as an important source of water that could be utilized for domestic purposes, drinking, agriculture, and in shallow aquifer feeding. Climate models on effectiveness of RWH systems in different climatic zones in South Africa, using daily rainfall time series data from 1988 to 2007, showed that the optimum RWH system could secure a 10–15% water saving in arid zones, 15–20% in both semi-arid and dry subhumid zones, and 30–40% in humid zones.

There have been a limited number of studies on the potential of RWH system in the arid regions of Australia, in contrast to urban areas of Australia where RWH systems have become popular in recent years due to government incentives and a greater environmental awareness in the community. In many remote arid regions of Australia, RWH is often the only viable potable water supply and is thus an important component of rural water provision. In spite of various limitations in the arid regions, RWH has a number of potential applications as a supplementary water supply to increase water security. The RWH system has potential applications for storm water management and rehabilitation of pastoral land. Studies on financial aspects of RWH systems have often produced conflicting results. Misrepresentation of operational costs of RWH systems in particular has led to misleading conclusions in many cases. The majority of research in Australia and other countries with strong economies has shown that water from RWH systems is generally more expensive than tap water at current water prices.

In view of the varying and conflicting results, there is a need to standardize the methods of economic analysis for RWH systems. However, it has been found that RWH systems in the UK are not likely to present any payback period and any research that finds they can should be thoroughly examined. It appears that the majority of researchers have found that RWH systems are not financially viable. Conflicting results may be affected by a number of financial assumptions and modeling parameters. There seems to be little research on the consumer price index and also the relationship between labor and material costs with respect to RWH systems, as most research has been done locally. Improper consideration of maintenance and operational costs is responsible for many of the conflicting conclusions on the economic viability of a RWH system. Ongoing maintenance expenses have often been identified as a primary reason for RWH system costs outweighing the benefits.

However, an additional recharge not only has beneficial impacts but in arid zones, the rise in groundwater level also often has negative impacts in the form of waterlogging and soil salinization, which require active drainage, especially in nearly horizontal plains. "Vertical drainage" by pumped drilled wells is practiced in such areas, for example in Pakistan and Uzbekistan.

5.2 Successes of Water Harvesting

There are many success stories of RWH to supply water for small irrigation areas, livestock, and artificial recharge of groundwater in developed and developing countries including Asia, Africa, Brazil, China, India, Latin America, USA, Japan, Germany, and Singapore. These case studies can help in determining future strategy for RWH to reduce water scarcity to some extent. Most of the success stories cover rooftop RWH and also surface storage structures like lakes and ponds. Gaps in water supply can be bridged to a limited extent through RWH, providing a strategy is put in place and the harvesting structures are well maintained. However, limited effort has been made to determine the effect of rainfall characteristics on overall RWH system performance.

5.2.1 America

In the United States, the natural annual recharge of groundwater is relatively small because of the slow rates of movement of groundwater and the limited opportunity for surface waters to penetrate the land surface. To supplement this small natural recharge, a recent trend for artificial recharge has been developing. Extensive spreading basin-type recharge facilities are common in California. The recharge rates in such basins vary from 0.1 to 2.88 m day^{-1}. In California and Colorado, channel methods have been used extensively. In California, after a period characterized by overabstraction from the Central Valley aquifer system (1950s and 1960s), several artificial recharge schemes and the partial substitution of groundwater abstraction by surface water transfers, complemented by the establishment of a market for water rights, have contributed to a partial improvement in the groundwater budget.

In Long Island, New York, since 1936 basins for putting storm water into the ground have been in operation, which have salvaged and recharged storm water into the aquifer. Recharge wells have been extensively used in Long Island and in coastal tracts of Los Angeles to build up and maintain a groundwater ridge to control sea water intrusion. Recirculation of water for cooling purposes is another application of the recharge well method. In recent years, wells have been used increasingly to store heat underground by recharging hot water into confined aquifers. Results of a recharge experiment carried out in 1946 at Camp Peary, Virginia, indicated that it is possible to store fresh water in saline water-bearing aquifers and extract it subsequently without significant losses.

Experiments carried out at Peoria, Illinois, to demonstrate the efficacy of the pit method to recharge a glacial drift aquifer comprising sand and gravel showed that a maximum infiltration rate of 60 m day^{-1} and mean rates of 12–31 m day^{-1} could be reached. Use of clean sand in the initial stages resulted in repeated clogging by silt carried by the water supplied from the Illinois River, with decrease in the recharge rate by as much as 60% in three months. The average recharge rate in a period of 146 days was 6.84 million L day^{-1}. Replacement of the fine sand filter by pea gravel gave satisfactory performance and increased the recharge rate to 8.36 million L $year^{-1}$. Experiments carried out on pits with different side slopes showed that pits with flatter side slopes provided a greater proportion of wet table surface area per unit of submergence, allowing higher rates of infiltration. Some wells have high recharge capacities, such as those at Louisville, Kentucky, and Orlando. In Florida, rates up to 6.5 and 40 million

L year^{-1} have been obtained. However, there are reports of recharge wells having lost 50% of their capacity within 1–10 weeks. Many successful recharge wells either receive filtered water or recirculate clean water.

In the United States, RWH systems for managing recharge to groundwater exist in every region of the country, with properly sited, designed, constructed, operated, and maintained managed aquifer recharge (MAR) projects. In 1955, over 700 million gallons of water per day were artificially recharged, although the total recharge volume was only about 1.5% of the groundwater withdrawn that year. The water was derived from natural surface sources and returns from air conditioning, industrial wastes, and municipal water supplies. Despite many successful MAR projects, additional research is needed on the long-term physical and chemical impacts on aquifers, optimal management in different environments, uses of storm water and reclaimed water, and institutional considerations. By early 2014, the Central Arizona Groundwater Replenishment District (CAGRD) had developed a diverse portfolio of short-, mid-, and long-term water supplies yielding in excess of 43 500 acre feet per year over the next 30 years, as part of the statutory obligation to replenish excess groundwater withdrawn by its members in the Phoenix, Pinal, and Tucson Active Management Areas. Replenishment may be accomplished through the operation of underground storage facilities or groundwater savings facilities, or through the purchase of long-term underground storage credits held by third parties.

Florida uses MAR extensively. The Peace River Manasota Regional Water Supply Authority supplies an average of 25 million gallons of water daily. The water, skimmed during higher flow periods from the Peace River, is treated and then injected into an aquifer, for later recovery. The process allows water to be withdrawn from the river during "wet" periods, stored underground, and then used during "dry" periods when river levels are low. In Boise, Idaho, United Water Resources uses MAR to inject and store seasonally available high-quality groundwater and treated surface water into aquifers of poorer quality water for use during peak demand season using municipal wells. New Jersey reportedly has the oldest US aquifer storage and recovery project. Operational since 1967, treated drinking water is injected using wells and stored underground during most of the year. During summer when local water demand increases dramatically, a significant portion of this demand is met by withdrawing water from the underground aquifer using the same wells.

The city of Rio Rancho, New Mexico, has demonstrated that surface infiltration and direct injection methods can be used to safely replenish underlying aquifers with a purified, reclaimed water source when the proper level of treatment is employed. The surface infiltration system has the capacity to recharge the underlying aquifer, present at a depth of nearly 500 ft below ground surface, at a rate of 650 000 gallons per day. The direct injection facility is capable of replenishing the same groundwater system at a rate of 1 million gallons per day. The city is currently constructing an advanced water treatment facility that uses an ozone-based advanced oxidation process, followed by a biologically activated carbon step to reduce remaining waste water-derived organics in the reclaimed water prior to direct injection. This process has been used extensively for drinking water treatment and has shown great potential for reclaimed waters when coupled with membrane bioreactor water reclamation systems. The Dayton Water Department in Ohio provides drinking water to approximately 440 000 people, using the infiltration lagoons and ponds throughout the well field to enhance natural groundwater recharge. In Oregon, Madison Farms uses artificial recharge as a way to collect high-turbidity winter floodwater. The water is then cleaned to potable standards, injected into the aquifer, and withdrawn in the summertime for irrigation use.

5.2.2 Australia

In many remote arid regions of Australia, RWH is often the only viable potable water supply and it is thus an important component of rural water provision. Yuen et al. (2001) reviewed the applications of various water harvesting techniques for four small communities: Jigalong, Cheeditha, Wittenoom, and Karrath in the Pilbara region of Western Australia. It was found that water harvesting could provide valuable irrigation water in harsh climatic and environmental conditions, and was environmentally sustainable, requiring relatively low maintenance. In South Australia, the most arid state, 51% of people use rainwater tanks and 80% of these are located in arid parts of the state. Furthermore, the RWH system has the potential for storm water management and pastoral land rehabilitation.

5.2.2.1 Infiltration Basins

Since the 1960s, MAR has been central to maintaining elevated water tables across the area and preventing coastal saline intrusion that would otherwise have occurred due to groundwater pumping (Charlesworth et al. 2002). Recharge of up to 45 gigaliters year^{-1} has been achieved over about 40 years using recharge pits situated over coarse sand deposits; by sand dams within the Burdekin River near Townsville that slowly release water from upstream storage; and by diversions to constructed channels and natural waterways. River water with the lowest turbidity levels is allowed to enter the recharge pits to maximize the time-span of operation between scraping and removal of the basin floor to remove deposited particles and renovate recharge rates. This is typically done at two-year intervals. Some of the pits are reported have a recharge capacity of up to 20 million liters day^{-1} (Marchant and Bristow 2007). About 38 000 ha of sugarcane and other crops rely on irrigation, including from shallow groundwater, which also serves as a drinking water supply.

5.2.2.2 Soil Aquifer Treatment for Water Reuse

Alice Springs, in the Northern Territory, relies on very slowly renewed groundwater reserves from deep aquifers for its water supply. To prevent winter overflows of sewage effluent to Ilparpa Swamp and to provide irrigation water supplies for a horticultural development, a feasibility study was conducted (Knapton et al. 2004), and a site was identified where 600 ML year^{-1} reclaimed water could be stored in a palaeochannel aquifer, and then recovered for irrigation supplies.

5.2.3 China

In RWH system in some arid regions of China, the water collected from rooftops immediately after rainfall showed good quality in relation to inorganic compounds in the rain water and the quality of water was in accordance with WHO drinking water standards (Yihdego 2015; Yihdego and Eslamian 2016).

5.2.4 Egypt

On the Sinai Peninsula in Egypt, RWH is an important source of water that could be utilized for domestic purposes, drinking, agriculture, and in shallow aquifer feeding. The groundwater system is recharged from rivers and canals, irrigation water infiltrated through the aquitard, and drainage of flash rains from east desert wadis. In the Nile Valley and delta, the relatively small difference in heads between two interconnected groundwater bodies, which consist of the water in the clay cap (aquitard, shallow subsoil water table) and the main semi-confined aquifer (piezometric head), causes movement of water in a vertical direction. The interaction between the two bodies also has implications for agricultural production. The groundwater reservoir loses water along the Rosetta and Western desert depressions. Some of the fresh water escapes to the north delta lagoons and directly or indirectly towards the Mediterranean Sea and Suez Canal, which form the north and east boundaries, where the sea water intrudes and the transition zone between fresh water and salt water depends on the dynamic conditions of the system.

5.2.5 Israel

Although primarily a desert country, Israel has emerged as a leader in harnessing its scant water resources, its efficient optimal use and effective management, through a network of canal systems which depend substantially on groundwater and transport water from the surplus in the northern and coastal regions to areas of deficit. The entire transportation of water is based on the principle of minimizing evaporation and seepage, with covered channels and piped water supply. Isotopes are used to estimate aquifer storage, recharge and mixing rates, and wells are monitored continuously to ensure no overexploitation of aquifers. Groundwater is not considered separate from surface water, and artificial recharge of groundwater is practiced widely. In Israel, hydrodynamic equilibrium has nearly been restored in the coastal aquifer by a combination of artificial recharge (90–135 million m^3 $year^{-1}$) and reduced groundwater withdrawal (by approximately 100 million m^3 $year^{-1}$, during the 1990s).

5.2.6 India

In India, the Central Ground Water Board has implemented more than 200 RWH and recharge schemes and assessed the impact of recharging on groundwater levels. The rise in groundwater levels was found to be in the range of 0.15 to 12 m in different parts of the country. Artificial recharge activities require group action at community or farm level with financial assistance from government and financial institutions. There is also a need to determine a viable public–private partnership model. The success of the program will also depend on interagency co-operation for joint programming, planning, and implementation.

5.2.6.1 Check Dams

These are an important form of water management and contribute considerably to societal welfare through providing protection from seasonal floods. Check dams, generally constructed for soil conservation, can also be considered as mini/micro-percolation tanks for augmenting groundwater. During unprecedented droughts, social workers and NGOs have undertaken a large number of water-harvesting projects to recharge groundwater. With the apparent success of these efforts, in 2000–2001, the Government of Gujarat under the Participatory Water Conservation Program invested over Rs 1180M for the construction of more than 10 000 check dams in Saurashtra, Kutch, Ahmedabad, and Sabarkantha regions, with 60% funding from the government and 40% from beneficiary stakeholders. The responsibility for managing the construction works remained with NGOs.

A socioeconomic cost–benefit analysis of over 100 check dams (which is about 1% of the total number of check dams in Gujarat), carried out in 2002, suggested that localized RWH systems of check dams in Saurashtra were an effective solution to water crisis, using local skill-based, cost-effective, and environment-friendly simple technologies (Shingi and Asopa 2002). The 60:40 scheme was economically sound, capable of attracting beneficiaries contributing their share through labor, equipment, and/or money in a highly participatory approach; women were the major beneficiaries of the alleviation of drinking water and livestock feed problems. The project did not replace or endanger human or wildlife habitat and prevented the entry of undesirable contractors into partnership with government. Groundwater recharge is often best accomplished by this approach. About 30–40% of applied irrigation water seeps from irrigation fields, a part of which recharges groundwater (CGWB 1995) and about 55–88% of water infiltrates from paddy fields.

5.2.6.2 Johads

The most important achievement is the successful runoff water harvesting and rejuvenation of the Arvari and Ruparel Rivers in Alwar district, Rajasthan. Johads built upstream of the main river transformed the ephemeral stream into a perennial river and reportedly helped to recharge aquifers. Increase in the shallow aquifer groundwater table was seen from ~100–120 cm depth to 3–13 m. In the villages where johads have been revived, water is shared among the villagers and the farmers are not allowed to grow water-intensive crops. The areas under single cropping and double cropping, which were previously 11% and 3%, increased to 70% and 50% respectively. The initial ~7% forest cover increased to 40% by agroforestry and social forestry, providing sufficient fuel wood. Fisheries also developed.

The biggest beneficiaries of RWH are relatively well-off farmers, who have the resources to spend on extracting water (e.g. tube wells, submersible pump-sets, etc.), as the main outcome of water harvesting is to recharge neighboring groundwater. Poorer villagers also get employment as waged laborers during the period of construction. Johads may be cheaper to build than concrete anicuts. The Embassy of Sweden contributed 16.46 MSEK (about Rs 116 million) as core funding from 2002 to 2009 (SIDA 2013). During the latest strategy period (2009–2013), the Embassy of Sweden supported such programs with 4.5 MSEK (about Rs 29 million) (SIDA 2013).

5.2.6.3 Tanks/Ponds

Artificial recharge through infiltration ponds can be applied almost anywhere, if there is a supply of clean fresh water available at least part of the year, the bottom of the pond is permeable, and the aquifer to be recharged is near the surface. In some areas, percolation tanks/ponds are constructed across or near streams to impound rainwater and retain it for longer to allow groundwater recharge. In some cases where aquifers are under stress, even in many traditionally managed tank irrigation systems originally built for surface supplies, when gravity-supplied water from the tank is insufficient for crop production, tanks are being converted into percolation tanks by closing the outlet sluices and allowing the stored tank water to recharge the wells in the command area (Sakthivadivel and Gomathinayagam 2004).

5.2.6.4 Floodwater Diversion

According to a study undertaken on the Lakhaoti branch canal, covering over 200 000 ha, bounded by the drainage canals of the Kali and Nim rivers, in the districts of Ghaziabad, Bulandsher and Aligarh in western Uttar Pradesh and the Madhya Ganga Canal Project located in the lower Ganga canal commands, during the *kharif* season, the diversion of surplus Ganga floodwater helped to raise the groundwater table from 6.6 to 12.0 m, and brought down the cost of pumping for irrigation (Chawla 2000). The irrigated area in the project area increased from 1251 ha in 1988–1989 to 35 798 ha in 1999–2000, and the area under paddy irrigation increased to 14 419 ha from 83 ha. The total annual cost of pumping for paddy cultivation due to canal seepage declined, resulting in cost saving for the project as a whole. Although, in 1987, the Government of India circulated a model Groundwater Regulation Bill among the states, none of the states adopted either this bill or its modified version. The government of Tamil Nadu has enacted a groundwater regulation act pertaining to Chennai metropolitan area to overcome the grave situation due to a severe drinking water crisis. The water table near the northern part of the city rose in 2001–2002 due to implementation of this act and other artificial recharge measures.

5.2.7 Oman

The hydrodynamic equilibrium of the groundwater systems in several areas has been completely or nearly restored with the help of large recharge dam schemes and a licensing system for drilling wells and abstracting groundwater.

5.2.8 Netherlands

Depletion of fresh groundwater lenses overlying saline water in the dune area along the west coast of The Netherlands has been effectively prevented for many years by means of artificial recharge.

5.2.9 Sweden

Pits constructed on the tops of eskers are used to recharge aquifers. The recharged water is extracted by wells located at the fringe of the aquifer. The rate of infiltration in the pits is about 1.5–15 m day^{-1}.

5.2.10 Russia

The recharge basin system is most common due to periodical flooding in river valleys, as well as periodical discharges of water from water storages. It is implemented into the largest Australian Guidelines for Water Recycling (AGWR) systems in Tbilisi, Riga, and Tashkent. Water harvesting from periodical flooding of rivers is another method in some systems in Kazakhstan and the Crimea.

5.3 Failures of Water Harvesting

5.3.1 Australia

There have been a limited number of studies on the potential of RWH systems in the arid regions of Australia in contrast to the urban areas where RWH systems have become popular in recent years due to government incentives and a greater environmental awareness in the community. Investigations on the feasibility of different types of RWH system at seven locations, in the arid regions of Western Australia, South Australia and the Northern Territory, suggest that these regions are predominantly groundwater reliant. However, many areas have issues with groundwater quality.

5.3.2 India

In most project areas, risks to the local environment and population due to disturbance of hill slopes, soil erosion, deforestation, and disruption of flora and fauna, associated with plans for construction of facilities, access roads, etc. are considered as failures of harnessing water from high rainfall surface runoff. Generally, scientists, managers, policy makers, and the public have inadequate knowledge, awareness, and different perceptions of vulnerability and risk. The primary disadvantages are large land area requirements, the possibility of recharging contaminated water or degrading groundwater quality, and challenge in identifying locations with surface and subsurface conditions amenable to infiltration. Identifying areas suitable for such projects and assessing their influence on groundwater levels and flows can be difficult due to limited data on complex surface and subsurface conditions. High levels of surface collection of water decrease the recharge of the aquifer (Datta and Goel 1977; Datta et al. 1973, 1980).

The impacts could vary with project location, size, and operating conditions. In some cases, the projects failed due to conflicts among new-traditionalist views and within premodern communities, and also because of local power relations in accessing natural resources. Such approaches result in water harvesting practices being incorporated into the wider practice of land grabbing, and may be intended by local politicians and other influencers to allow continued and extended accumulation of land at the expense of local communities, their livelihoods, and their relationship with the area.

5.3.2.1 Tanks/Ponds

Recharge through infiltration ponds can be useful, provided there is a supply of clean fresh water available at least part of the year, the bottom of the pond is permeable, and the aquifer to be recharged is at or near the surface. Most evaluation studies on percolation tanks are qualitative, based on the response of the aquifer system or increase in crop yield. Although over 250 000 tanks and ponds exist in hard-rock areas of peninsular India, as most of the tanks are old their storage capacity has reduced due to siltation, thereby reducing the recharge volume of water considerably. Studies on nine percolation tanks in the semi-arid regions of the Noyyal, Ponani and Vattamalai river basins in Tamil Nadu and Kerala showed that percolation rates were reduced due to the accumulation of silt (Raju 1998).

In some cases, using tank command areas for multiseason cropping with higher intensities than the system was originally designed to meet resulted in more and more farmers overextracting groundwater through wells, leading to water table decline, especially in areas where density of wells is high and rainfall is moderate to low. In Punjab, artificial recharge studies in the Ghagger River basin using injection wells and canal water as the main surface water source showed that the recharge rate from pressure injection was 10 times that of gravity systems and that maintenance was required to preserve efficiency (Muralidharan and Athavale 1998). Evaluation studies on percolation tanks in southern peninsular India indicate that recharge efficiency varies from 30% to 60%, depending upon the hydrogeological situation. The effectiveness of induced recharge methods depends on the proximity of surface water bodies, hydraulic conductivity of the aquifer, area and permeability of the stream/lake bed, and hydraulic gradient created by pumping. Indirect methods generally provide less control over quantity and quality of the water than do direct methods.

5.3.2.2 Johads

Although NGOs with local participation have made great efforts to rejuvenate defunct johads and construct new ones on tributary streams, with indications of water level rise in adjacent wells and facilitating irrigation on cultivated lands, from the reported literature it is difficult to establish a direct link between increase in water level and creation of johads. Potential recharge from these structures reaches a limiting point with increasing rainfall, which is a factor of the maximum storage capacity in the structure. Also, bandhs were found to have an average potential recharge of 37.5 mm day^{-1} and Johads 17.8 mm day^{-1}, thus highlighting the importance of careful positioning of the structures. The demand for anicuts is higher in villages due to their durability and potential in terms of irrigation development. However, to generate funds, NGOs seem to refer to all their water harvesting structures by the generic name "johad" to maintain their identity as grassroots organizations promoting "local traditional knowledge" and community participation.

Analysis of the well data is difficult but taking the most unaffected sections of the time series and assuming a constant specific yield, estimates of groundwater recharge were 7.2–11.3 mm day^{-1}, increasing down the catchment in elevation. The recharge values from the wells were much lower than the estimates of potential recharge from the RWHS. This suggests that either the recharge from RWH is not reaching the aquifer or the aquifer has large transmissivity and therefore strong lateral flow (SIDA 2013). About 30% potential recharge remains stored in the soil while at least another 30% moves laterally. As a result, the water recharged gets spread over a larger area, depending on the size of the aquifer. Moreover, little direct support for these activities have arisen other than from the Embassy of Sweden, because of the challenging position sometimes taken by the Government of India with regard to land use and communities (SIDA 2013). However, the substantially reduced amount of funding support from the Embassy of Sweden during 2009–2013 (almost one-fourth of the funding in 2002–2009) (SIDA 2013) was a matter of concern.

The Alwar district (where the johads are reportedly successful) is situated in north-east Rajasthan, covering 8380 km^2 area (~2.45% of the total area of the state). The district has 2226 villages and receives 577.7 mm annual rainfall on average. There is no natural lake in the district, and the Ruparel and Sabi are the two ephemeral rivers. Nearly 83% of the groundwater in the district is considered as fresh to moderately saline with electrical conductivity <4000 s cm^{-1}. The surface water quality is low at Alwar and its surroundings. The quality has deteriorated because there is no proper sewerage system and 5 MLD of highly polluted effluent is discharged from industrial areas into drains, which finally flows towards Mansarovar. A large quantity of effluent generated by Alwar city is discharged without any treatment. The groundwater fluoride was found to be 0.45–3.6 ppm in various parts of the Alwar region, which is more than the permissible limit in most of the samples studied. The quality of drinking water is very poor, and can be used for drinking and cooking only after prior treatment.

A large number of organic chemicals can be present in natural catchment waters, urban storm water, reclaimed water and treated water for drinking, which include pesticides, hydrocarbons, polycyclic aromatic hydrocarbons, disinfection by-products and emerging chemicals of concern such as endocrine-disrupting chemicals, personal care products, pharmaceuticals, and flame-retardants. Hence it is quite possible that some trace organic substances can discharge to streams or lakes if they are in close proximity to the recharge facility. Also, a large number of organisms require protection, and neither their dose–response to each of the chemicals, nor the effects of interactions between these substances and the compounding effects of other environmental stresses (e.g. water/salinity stress) are known or will be known in the near future. In Jodhpur, many nadis have been severely polluted because of poor maintenance and negligence, destruction of catchment areas, and uncontrolled urbanization (Mishra 2001).

In Saurashtra district, Gujarat, in spite of the poor potential offered by low rainfalls, high variability and high evaporation rates, poor storage capacity or specific yield of the basalt formations, significant recharge efforts have been made. Around 20 000 check dams built to capture rainwater and recharge the aquifers are able to store only a small fraction of the surplus runoff. During good rainfall years, the aquifers become saturated with natural recharge immediately after the rains, leaving no space for entry of water from the recharge systems (Muralidharan and Athavale 1998; Shah 2002).

5.3.3 Basin-Level Impacts

In most water-scarce regions, there are complexities in the economic evaluation of RWH, due to lack of reliable scientific data on beneficiaries, water inflows, runoff collection and storage efficiency, value of the incremental benefits generated, and scale considerations. Therefore, in RWH programs, there is little or no emphasis on the potential local water supplies and the demand to cater for local low water supply potential; water demand far exceeds the supply potential. Also, with higher degrees of basin development, the marginal benefit from water harvesting at the basin level reduces, while marginal cost increases. In many basins, there is a strong "tradeoff" between maximizing hydrological benefits and improving cost-effectiveness. Many water-scarce basins are characterized by wide disparity in demand between upper catchments and lower catchments, so that there is a tradeoff in maximizing benefits of upstream water harvesting with optimizing basin-wide benefits. In many water-scarce basins, local water harvesting merely divides the hydrological benefits rather than augmenting them.

While the impact of upstream artificial recharge on downstream users in basins may differ from basin to basin, all water uses are likely to have some impact on users elsewhere in the system, and this is greatest when all available supplies in the basins are fully allocated, and in cases with marked inter- and intraannual variation in rainfall (Molden and Sakthivadivel 1999). For instance, the Ajil reservoir, which supplies water to the city of Rajkot, located downstream of the watershed in the Saurashtra region of Gujarat, is considered as one such water-scarce basin, with high variation in rainfall (200–1100 mm $year^{-1}$) and runoff coefficient. Analysis of the rainfall and inflow data (1968–2000) to the reservoir indicated that while rainfall remained more or less the same throughout, the runoff coefficient declined to almost half of its original value, especially after 1985 when flow to the reservoir began to decline sharply due to the construction of thousands of check dams and percolation ponds within the watershed.

The indirect recharge method also had a negative impact in managing large irrigation systems due to siphoning of considerable quantities of water to areas not included in the command area plan (Neelakantan 2003). Moreover, when indirect methods of recharge and retrieval are practiced, the water recovered consists of a small fraction of groundwater while a larger fraction of abstracted water comes from rivers or lakes. Since the surface water–groundwater system is interconnected, when groundwater is depleted stream flow depletion takes place, which also affects aquatic habitats. In areas adjacent to the Kali River in western Uttar Pradesh, groundwater contamination due to river water contamination further reaffirms the surface and groundwater interaction and indicates the need for holistic action for water resources management.

In Wakal Basin, southern Rajasthan, the $\delta^{18}O$ and $\delta^{2}H$ in groundwater suggested that while artificial recharge was present in many of the wells sampled within the proximity of the RWH structures at two study sites, the proportion of artificial recharge in these wells was 25–75%. Artificial recharge proportions (52–83%) were also recorded in wells downstream of an Indian RWH structure situated in a similar hard-rock terrain (Sukhija et al. 2005; Stiefel et al. 2009). The use of $\delta^{18}O$ revealed that 50–70% of a groundwater system in Kenya originated from analogous impounded surface water from a nearby lake (Ojiambo et al. 2001). However, the data also suggested that artificial recharge was not present in every well. Spatial trends in the tracer data revealed that the wells located upstream of the RWH structures generally were not influenced by artificial recharge. This finding supports the regional downstream groundwater flow direction evident in the water table data in other RWH studies (Sukhija et al. 2005; Stiefel et al. 2009). Studies based on geographic information system (GIS) modeling in part of the Pajaro Valley Groundwater Basin, California, suggest that only about 7% (15 km^2) of the basin may be suitable for MAR, and it is necessary to determine the variation of impacts of MAR with project location, size, and operating conditions.

Exceptions to the general spatial trend of decreasing artificial recharge in the downstream direction are also present within the study sites, suggesting that local groundwater flow paths are more complex. The mixing estimates reveal that wells located relatively closer to the source of artificial recharge did not always receive higher proportions of artificially recharged groundwater. The slightly depressed water table at some locations suggested that there might be subregions of groundwater that do not readily mix with surrounding groundwater. These abnormalities evident in the tracer data and water table are partially explained by characteristics of the hydrogeology of the study area. Since groundwater flow is restricted to the fractures present in the hard-rock aquifers, the structural geology of the study sites provides further insight into the movement of groundwater. The data further suggest that a preferential flow recharge process is applicable to artificially recharged water from the RWH structures, as well as natural recharge.

References

Central Ground Water Board (CGWB) (1995). Report on Groundwater Resources of India. New Delhi: Government of India.

Charlesworth, P.B., Narayan, K.A., Bristow, K.L., Lowis, B., Laidlow, G. and McGowan, R. (2002) The Burdekin Delta – Australia's oldest artificial recharge scheme. In: Management of Aquifer Recharge for Sustainability, Proceedings of the Fourth International Symposium on Artificial Recharge (ISAR4), September 22–26, 2002, Adelaide, pp. 347–352.

Chawla, A.S. (2000). Ground Water Recharge Studies in Madhya Ganga Canal Project. Colombo: Consultancy Report to IWMI.

Datta, P.S., Desai, B.I., and Gupta, S.K. (1980). Hydrological investigations in Sabarmati basin – I. groundwater recharge estimation using tritium tagging method. Proceedings of the Indian National Science Academy Physical Sciences 46 (1): 84–98.

Datta, P.S. and Goel, P.S. (1977). Groundwater recharge in Punjab state (India) using tritium tracer. Nordic Hydrology 8: 225–236.

Datta, P.S., Goel, P.S., and Rama Sangal, S.P. (1973). Groundwater recharge in western Uttar Pradesh. Proceedings in Mathematical Sciences 78 (A): 1–12.

Knapton, A., Jolly, P., Pavelic, P. et al. (2004) Feasibility of a Pilot 600 ML/yr Soil Aquifer Treatment Plant at the Arid Zone Research Institute. Department of Infrastructure, Planning and Environment, Alice Springs.

Marchant, S. and Bristow, K.L. (2007) Burdekin field trip: an awesome 'dam' experience. CSIRO Land and Water Science Report.

Mishra, A. (2001). The Radiant Raindrops of Rajasthan. Dehradun: Research Foundation for Science, Technology and Ecology.

Molden, D. and Sakthivadivel, R. (1999). Water accounting to assess use and productivity of water. International Journal of Water Resources Development 15 (1 & 2): 55–71.

Muralidharan, D. and Athavale, R.N. (1998). Base Paper on Artificial Recharge in India. Hyderabad, India: National Geophysical Research Institute, CSRI.

Neelakantan, S. (2003) A Gossipmonger's Revisit to Chettipalayam. Working Paper No. 142. Madras Institute of Development Studies, Chennai.

Ojiambo, B.S., Poreda, R., and Lyons, W.B. (2001). Ground water/surface water interactions in Lake Naivasha, Kenya, using δ18O, δD, and 3H/3He age-dating. Ground Water 39 (4): 526–533.

Raju, K.C.B. (1998). Importance of recharging depleted aquifers: state of the art of artificial recharge in India. Journal of Geological Society of India 51: 429–454.

Sakthivadivel, R. and Gomathinayagam, P. (2004) Case Studies of Locally Managed Tank Systems in Karnataka, Andhra Pradesh, Gujarat, Madhya Pradesh, Gujarat, Orissa, and Maharashtra. Report submitted to IWMI-TATA Policy Programme, Anand.

Shah, T. 2002. Decentralized Water Harvesting and Groundwater Recharge: Can These Save Saurastra and Kutch from Desiccation? IWMI-TATA Water Policy Research Program Annual Partners' Meet, Gujarat.

Shingi, P.M. and Asopa, V.N. (2002). Independent Evaluation of Check Dams in Gujarat: Strategies and Impacts. Ahmedabad: Centre for Management in Agriculture, Indian Institute of Management.

Stiefel, J.M., Melesse Assefa, M., McClain Michael, E. et al. (2009). Effects of rainwater-harvesting-induced artificial recharge on the groundwater of wells in Rajasthan, India. Hydrogeology Journal 17: 2061–2073.

Sukhija, B.S., Reddy, D.V., Nagabhushanam, P., and Nandakumar, M.V. (2005). Efficacy of percolation ponds as artificial recharge structures and the controlling factors. Journal of the Geological Society of India 66: 95–104.

Swedish International Development Cooperation Agency (SIDA) (2013) SIDA Decentralised Evaluation 2013:28. Available at: www.sida.se/publications.

Yihdego, Y. (2015). Water reuse in hilly urban areas. In: Urban Water Reuse Handbook (ed. S. Eslamian), 903–913. Boca Raton: CRC Press.

Yihdego, Y. and Eslamian, S. (2016). Drought management initiatives and objectives. In: Handbook of Drought and Water Scarcity. Vol. 3: Management of Drought and Water Scarcity (ed. S. Eslamian and F.A. Eslamian). Boca Raton: CRC Press.

Further Reading

Amos, C.C., Rahman, A., and Gathenya, J.M. (2016). Economic analysis and feasibility of rainwater harvesting systems in urban and peri-urban environments: a review of the global situation with a special focus on Australia and Kenya. Water 8: 149.

Datta P.S. (2014). Need for Better Groundwater Governance for Sustained Water Supply. IPHE National Conference on Piped Water Supply and Sewerage Systems, January 24–25, New Delhi.

Datta P.S. (2015). Traditional Rainwater Harvesting at Crossroads – Time to Seek Ethical Opportunities for Reform. Proceedings of the Aqua Foundation IX World Aqua International Congress, Reviving Traditional Water & Environment Conservation Techniques, November 26–27, New Delhi, pp. 93–115.

Gupta, S. (2011). Demystifying 'Tradition': the politics of rainwater harvesting in rural Rajasthan, India. Water Alternatives 4 (3): 347–364.

Imteaz, M.A., Ahsan, A., Naser, J., and Rahman, A. (2011). Reliability analysis of rainwater tanks in Melbourne using daily water balance model. Resources, Conservation and Recycling 56: 80–86.

Mudgal, K.D., Kumari, M., and Sharma, D.K. (2009). Hydrochemical analysis of drinking water quality of Alwar District, Rajasthan. Nature and Science 7 (2): 30–39.

Parthasarathi, G.S. and Patel, A.S. (1997). Groundwater Recharge through People's Participation in Jamnagar Region, 51–56. Nagpur: Indian Water Works Association.

Rahman, A., Dbais, J., and Imteaz, M. (2010). Sustainability of rainwater harvesting systems in multistory residential buildings. American Journal of Engineering and Applied Sciences 3 (1): 889–898.

Rahman, A., Keane, J., and Imteaz, M.A. (2012). Rainwater harvesting in greater Sydney: water savings, reliability and economic benefits. Resources, Conservation and Recycling 61: 16–21.

Roebuck, R., Oltean-Dumbrava, C., and Tait, S. (2011). Whole life cost performance of domestic rainwater harvesting systems in the United Kingdom. Water and Environment Journal 25: 355–365.

Roebuck, R.M., Oltean-Dumbrava, C., and Tait, S. (2012). Can simplified design methods for domestic rainwater harvesting systems produce realistic water-saving and financial predictions? Water and Environment Journal 26: 352–360.

Shah, T. (1998). The Deepening Divide: Diverse Responses to the Challenge of Ground Water Depletion in Gujarat. Anand: Policy School.

Shah, T. (2008). India's master plan for groundwater recharge: an assessment and some suggestions for revision. Economic and Political Weekly 43 (51): 41–49.

Sinha, S., Gururani, S., and Greenberg, B. (1997). The "new traditionalist" discourse of Indian environmentalism. Journal of Peasant Studies 24 (3): 65–99.

Yang, S.-Q. (2015). Coastal reservoir – a technology that may dominate future water supply. Journal of Water Resource and Hydraulic Engineering 4 (4): 388–397.

6

SWOT Analysis of Water Harvesting Systems

CHAPTER MENU

6.1 Strengths of WHS – Local-Level Benefits

Water harvesting for artificial recharge of groundwater by spreading has important advantages over the injection method. Spreading methods, however, are useful only for recharge of unconfined aquifers. The infiltration is almost like natural rainfall infiltration. Normal storm runoff or primarily treated drainage waters (removal of harmful chemical constituents and suspended materials) can be utilized for artificial recharge. Cleaning of the infiltrating surface of the recharge structure, once in a while, is an easy process as it involves only scraping of the surface. Recharge by injection is the only method for artificial recharge of confined or deep-seated aquifers with poorly permeable overburden. The recharge is instantaneous and there are no transit and evaporation losses. The injection method is also very effective in case of highly fractured hard rocks and karstic limestones. In the recharge well injection method, as pumping wells are self-cleaning, dual-purpose injection (that is, injection via pumping) wells are more efficient.

6.1.1 MacroWH

The main benefits of macroWH are that it is simple to design and control, cheap to install (and to adapt) by individual farmers, and therefore easily replicable. There is the possibility of increased year-round water availability for domestic purposes, supplementary irrigation and livestock consumption. Risk of crop production failure is reduced by bridging prolonged dry periods. There is enhanced crop, fodder and tree production, and improved water use efficiency. Runoff collection efficiency is greater than with medium- or large-scale water harvesting systems. There are negligible conveyance losses and reduced damage from soil erosion, trapping of nutrient-rich

Water Harvesting for Groundwater Management: Issues, Perspectives, Scope, and Challenges,
First Edition. Partha Sarathi Datta.

sediments in runoff and flooding by storing excess runoff water. The area to be prepared for planting as well as fertilizer inputs are reduced compared to conventional preparation of the entire field.

6.1.2 Rock Catchment Masonry Dams

A major benefit of these systems is lower water loss through seepage. Water collected in rock catchment dams is often extracted for domestic and livestock consumption or supplementary irrigation.

6.1.3 Sand Dams

The sand reduces evaporation and contamination of the water in the sand body behind the dam, rendering the water suitable for livestock, domestic supply or small-scale irrigation.

6.1.4 Percolation Dams

These systems do not block groundwater flow, and serve three purposes: (i) reduce the speed of surface flow; (ii) increase percolation for the recharge of shallow aquifers; and (iii) obstruct the flow of sediments.

6.1.5 Check Dams

Check dams, generally constructed for soil conservation, are an important form of water management; and can be also considered mini/micro-percolation tanks for augmenting groundwater during unprecedented water crisis and droughts; they contribute considerably to societal welfare by providing protection from seasonal floods. Social workers and NGOs have undertaken a large number of water harvesting projects to recharge groundwater. Due to the apparent success of these efforts, in 2000–2001, the Government of Gujarat under the Participatory Water Conservation Program invested in the construction of more than 10 000 check dams in Saurashtra, Kutch, Ahmedabad and Sabarkantha regions, with funding from the government and beneficiary stakeholders.

An overall socioeconomic cost–benefit analysis of about 1% of the total number of check dams in Gujarat suggested that localized rainwater harvesting systems in Saurashtra were an effective solution to water crisis, using local skill-based, cost-effective, and environment-friendly simple technologies (Shingi and Asopa 2002). The scheme attracted donor investment via local peoplecontributing their share through labor, equipment and/or money; women were the major beneficiaries of the alleviation of drinking water and livestock feed problems. The project did not replace or endanger human or wildlife habitat, and prevented the entry of undesirable contractors into partnership with government. Applied irrigation water partly seeps from fields and recharges groundwater (CGWB 1995).

6.1.6 Johads

The most important achievement of these sysems is successful runoff water harvesting and rejuvenation of the Arvari and Ruparel Rivers in Alwar district, Rajasthan (Sharma

2006; Das 2010). Johads upstream of the main river transformed the ephemeral stream into a perennial river and reportedly helped to recharge aquifers (Borthakur 2008). Increase in the shallow aquifer groundwater table was seen. In villages where johads have been revived, water is shared among the villagers. The area under single cropping and double cropping increased. The original forest cover was increased by agroforestry and social forestry, providing more wood for fuel. The biggest beneficiaries were relatively well-off farmers, who have the resources to spend on extracting water (e.g. tube wells, submersible pump-sets, etc.), as the main outcome of water harvesting is to recharge neighboring groundwater. Poorer villagers also get employment as waged laborers during construction. Johads were cheaper to build than concrete anicuts. Fisheries were also developed.

6.1.7 Tanks/Ponds

In some areas, surface storage facilities (percolation tanks/ponds) are constructed across or near streams to impound rainwater and retain it for longer to allow groundwater recharge. In some cases where aquifers are under stress, in many traditionally managed tank irrigation systems originally built for surface supplies, when gravity-supplied water from the tank is insufficient for crop production, tanks are being converted into percolation tanks by closing the outlet sluices and allowing the stored tank water to recharge the wells in the command area (Sakthivadivel and Gomathinayagam 2004).

6.1.8 Floodwater Diversion

According to a study undertaken on the Lakhaoti branch canal, bounded by the drainage canals of the Kali and Nim rivers, in the districts of Ghaziabad, Bulandsher and Aligarh in western Uttar Pradesh of the Madhya Ganga Canal Project located in the lower Ganga canal commands, during the *kharif* season, the diversion of surplus Ganga floodwater helped to raise the groundwater table, and brought down the cost of pumping for irrigation (Chawla 2000). The irrigated area and the area under paddy irrigation increased. The total annual cost of pumping for paddy cultivation due to canal seepage declined, saving costs for the project as a whole.

6.1.9 Sunken Streambed Structures (Dohs)

Dohs are a low-cost alternative method of increasing groundwater in a semi-arid area where production of high-value legumes depends on irrigation. They represent the best way of expanding the extent of irrigated land, and bringing irrigation to more families. Small, multiple recharge points exist for replenishing groundwater from wells.

6.1.10 Recharge Wells

Worldwide, to enhance groundwater replenishment artificially, the main methods are recharge basins or recharge wells. Relatively high rates of recharge can be attained by this method.

6.1.11 Managed Aquifer Recharge

Managed aquifer recharge (MAR) is at the cutting edge of integrated water management, presenting opportunities for conjunctive management of surface water and groundwater resources and producing fit-for-purpose water supplies. MAR can help to sustain groundwater supplies and dependent ecosystems in heavily used aquifers or as an adaptation to climate change if environmental flows and downstream entitlements can be assured.

6.2 Weaknesses of WHS – Negative Impacts

It is commonly assumed that these rainwater harvesting (RWH) practices artificially recharge local groundwater and increase the water level in wells located near these structures, and this suspected increase in the water level of traditional rural wells would buffer seasonal and interannual declines in groundwater, thereby improving access to reliable water supplies. All over the world, including India, substantial investment has been made in rainwater harvesting, but by and large, these efforts lack hydrological planning and sound economic analysis. Research on the impact of local water harvesting on groundwater recharge activities is very sparse. Although substantial RWH efforts have been carried out in Rajasthan, very few studies have systematically investigated their actual artificial recharge potential in a scientific manner. Therefore, many RWH structures are built without a clear understanding of their effect on the local groundwater system. Further, much of the investment made into RWH cannot be properly scrutinized.

The primary disadvantages of RWH systems by water spreading artificially for managing recharge to groundwater include relatively large land area requirements, the possibility of recharging contaminated water or degrading groundwater geochemistry, and the challenge in identifying locations having surface and subsurface conditions amenable to infiltration. Artificial recharge by spreading has the disadvantage of loss of surface waters by evaporation and decrease in infiltration capacity of spreading structures by deposition of silt from the source water, deposition of dust from the atmosphere, and growth of vegetation in the infiltration structures. The very meager downward leakage through the aquitards and slow subsurface lateral movement of the groundwater make it uneconomical to artificially recharge the confined or semi-confined aquifers by either downward infiltration or lateral infiltration through the recharge zone. Identifying areas suitable for such projects and assessing their influence on groundwater levels and flows can be difficult due to limited data on complex surface and subsurface conditions and water flows.

In most such areas, to harness water from high rainfall surface runoff, plans for construction of project facilities, access roads, etc. are made assuming minimum risks from deforestation, disturbance to hill slopes, soil erosion, and disruption of flora and fauna. However, the environment and the population remain at risk. The spreading method, which involves stagnation of surface water, sometimes creates environmental problems. It is, therefore, necessary to adjust supply of water to these structures in such a way that large pools of stagnant water are not produced. The recharge rates also need to be controlled as higher groundwater mounds reduce infiltration rates. Generally, the public, scientists, managers, and policy makers have inadequate knowledge and awareness, and different perceptions of vulnerability and risks. Higher levels of rainwater collection decrease the recharge of the aquifer (Datta and Goel 1977; Datta et al. 1973, 1980).

In most water-scarce regions, generally, there is little or no emphasis on the potential local water supplies and the demand to cater for local low water supply potential, and water demand far exceeds the supply potential. Also, there are complexities in the economic evaluation of RWH, due to lack of reliable scientific data on beneficiaries, water inflows, runoff collection and storage efficiency, value of the incremental benefits generated, and scale considerations. With higher degrees of basin development, the marginal benefit from water harvesting at the basin level reduces, while marginal cost increases. In many basins, there is a strong "tradeoff" between maximizing hydrological benefits and improving cost-effectiveness. Many water-scarce basins are characterized by wide disparity in demand between upper catchments and lower catchments, so that there is a tradeoff in maximizing benefits of upstream water harvesting with optimizing basin-wide benefits. In many water-scarce basins, local water harvesting merely divides the hydrological benefits rather than augmenting them. Limited or poor integration between surface water and groundwater systems, and lack of inclusion of natural recharge, ultimately leads to reduction in potential for artificial recharge in hard-rock areas.

Soil infiltration capacity can be a limiting factor for recharge. In sandy and sandy loam soils, the infiltration capacity of the recharge area can be sustained through continuous removal of soils, but clay soils have inherent limitations. Results obtained from short-term infiltration tests carried out in different soil conditions show that the infiltration rate becomes negligible within 10 minutes of starting the test in the case of silty clay, whereas infiltration stabilizes within the first 25 minutes in the case of sandy loam. If the infiltration rate approaches to zero fast, this will negatively affect the recharge efficiency of percolation ponds. The extent of the problem would be larger in hard-rock areas (ideal for percolation ponds) with thin soil cover. Based on several infiltration studies, the rate of infiltration declines to a minimum value within 4–5 days of ponding. This will also have adverse effects on the performance of structures built in areas experiencing flash floods and high evaporation rates, solutions for which would be wetting or drying of pond beds through regulation of inflows.

In hilly catchments, due to relatively steeper slopes, high rainfall and runoff coefficients, the surface runoff water potential available for harvesting is generally high, and the area available for cultivation and agricultural water demand is also generally very low. In the valleys and plains, due to gentle slopes, lower rainfall, and low runoff coefficients, high potential evapotranspiration (PET) and deeper soil profiles, the surface runoff water potential for harnessing is generally low, but the area available for cultivation and agricultural water demand increase. In addition, the impacts could vary with project location, size, and operating conditions. Some of these questions can be resolved through field testing, but small-scale pilot field studies can be expensive and may provide limited spatial information.

New-traditionalist views do not appreciate the conflicts within premodern communities and are blind to local power relations in accessing natural resources. This may lead to water harvesting practices being incorporated into the wider practice of land grabbing, which may be encouraged by local politicians to allow continued and extended accumulation of land at the expense of local communities, their livelihoods, and their relationship with the area. The new-traditionalist views community as a homogeneous entity as opposed to the centralized state but fails to analyze relations among different social groups within a given community as well as the highly heterogeneous nature of the state.

Although injection is the only method for artificial recharge of confined or deep-seated aquifers with poorly permeable overburden, extreme purity of the source water as well as compatibility of source water and water from the aquifer to be injected are the necessary prerequisites. Otherwise, frequent clogging of injection structures by bacterial growth, chemical precipitation or deposition of silt results in heavy expenditure on well cleaning.

6.2.1 MacroWH

The main weakness of macroWH is that the catchment uses potentially arable land (with the exception of steep slopes). The catchment area has to be maintained, that is, kept free of vegetation, but crusts often develop on bare surfaces and therefore naturally reduce weed growth. Like all water harvesting systems, these systems can be damaged during exceptionally heavy rainstorms. Open and shallow rainwater ponds and dams may dry out after the rainy seasons, as the water is lost via evaporation and seepage (except for rock catchment and sand dams). Other disadvantages are that open storage structures can be contaminated by animals with the possibility of creating a breeding ground for disease-carrying insects. Sand dams are often contaminated as they are seldom protected from animals. In addition, since macroWH systems operate within a watershed scale, important issues that must be addressed include ownership, local institutions, and land/resource tenure. If maintenance is inadequate, soil erosion can occur and initial investment will be lost. Although such methods can increase water availability, risks still exist and in years with extremely low rainfall, they cannot compensate for overall water shortages. Several consecutive drought years always pose a problem, depending on the size of the storage system.

6.2.2 Check Dams

The gully area needs to be well protected against further erosion, otherwise the dam will be filled with sediments and converted into a productive gully. In the Saurashtra region, in spite of the poor potential offered by low rainfalls, high variability, and high evaporation rates, significant recharge efforts have been made. However, the biggest constraint in storing water underground during high rainfall years is poor storage capacity or specific yield of the basalt formations. During good rainfall years, the aquifers become saturated with natural recharge immediately after the rains, leaving no space for entry of water from the recharge systems. An estimated 20 000 check dams built in the region to capture the rainwater and recharge the aquifers are able to store only a small fraction of the surplus runoff.

6.2.3 Dohs

Group maintenance is required. Villagers may prefer larger and deeper "tanks." Dohs have limited capacity and dry up quickly, as do the wells.

6.2.4 Recharge Wells

To avoid clogging of the well, regular maintenance is needed. Clogging of the filter is one of the major problems to be solved. Other weaknesses include malfunction of the well due to aquifer geometry and characteristics, and retaining water for downstream users.

6.2.5 Tanks/Ponds

Most evaluation studies on percolation tanks are qualitative, based on the increase in crop yield. Although over 250 000 tanks and ponds exist in peninsular India hard-rock areas, as most of them are old their storage capacity has reduced due to siltation, thereby reducing the recharge volume of water considerably. Studies on nine percolation tanks in the semi-arid river basins in Tamil Nadu and Kerala showed that percolation rates were reduced due to the accumulation of silt (Raju 1998). In some cases, turning the tank command areas to multiseason cropping, with higher intensities than the RWH systems were originally designed to meet, resulted in farmers overextracting groundwater through wells, leading to water table decline, especially in areas where density of wells is high and rainfall is moderate to low.

In Punjab, artificial recharge studies in the Ghagger River basin using injection wells and canal water as the primary surface water source showed that the recharge rate from pressure injection was 10 times that of gravity systems and that maintenance was required to preserve efficiency (Muralidharan and Athavale 1998). The evaluation studies on percolation tanks in southern peninsular India indicate that recharge efficiency varies depending upon the hydrogeological situation. The effectiveness of induced recharge methods depends on the proximity of surface water bodies, hydraulic conductivity of the aquifer, area and permeability of the stream/lake bed, and hydraulic gradient created by pumping. Indirect methods generally provide less control over quantity and quality of the water.

6.2.6 Johads

Although NGOs with local participation have made great efforts to rejuvenate defunct johads and construct new ones on tributary streams, with indications of a rise in water level in adjacent wells and facilitation of irrigation on cultivated lands, from the reported literature it is difficult to establish a direct link between the increase in water level and construction of johads. Potential recharge from these structures reaches a limiting point with increasing rainfall, which is a factor of the maximum storage capacity in the structure. Also, bandhs were found to have on average a better potential recharge than johads, and this highlights the importance of careful positioning of the structures. The demand for anicuts is higher in villages due to their durability and potential in terms of irrigation development. However, to generate funds, NGOs seem to refer to all their WH structures by the generic name "johad" to maintain their identity as grassroots organizations promoting "local traditional knowledge" and community participation.

The analysis of well data was difficult because of the pumping but by taking the most unaffected sections of the time series and assuming a constant specific yield, estimates of groundwater recharge showed an increase down the catchment in elevation.

The recharge from the wells was much lower than the estimates of potential recharge from the RWHS. This suggests that either the recharge from RWH is not reaching the aquifer or the aquifer has large transmissivity and therefore strong lateral flow (SIDA 2013). Significant recharge remains stored in the soil while another significant part moves laterally. Thus, the water recharged gets spread over a larger area, depending on the size of the aquifer.

The Alwar district (where the hohads are reportedly successful) is situated in north-east Rajasthan, covering only ~2.45% of the total area of the state. There is no natural lake in the district, and the Ruparel and Sabi are the two ephemeral rivers. A large part of the groundwater in the district is considered as fresh to moderately saline with high electrical conductivity. The surface water quality is low at Alwar and its surroundings. The quality has deteriorated because there is no proper sewerage system in Alwar. A large quantity of effluent generated by the industrial areas is discharged into drains without any treatment. The groundwater fluoride is very high in various parts of Alwar region, more than the permissible limit in most of the samples studied. The quality of drinking water is very poor, and can be used for drinking and cooking only after prior treatment.

A large number of organic chemicals can be present in natural catchment waters, urban storm water, reclaimed water and treated water for drinking, including pesticides, hydrocarbons, polycyclic aromatic hydrocarbons, disinfection by-products and emerging chemicals of concern such as endocrine-disrupting chemicals, personal care products, pharmaceuticals, and flame retardants. Hence, it is quite possible that some trace organic substances will be discharged to streams or lakes if they are in close proximity to the recharge facility, and especially if cyclic recharge and recovery are not in annual balance. Also, a large number of organisms require protection, and neither their dose–response to each of the chemicals, nor the effects of interactions between these substances and the compounding effects of other environmental stresses (e.g. water or salinity stress) are known or will be known in the near future. In Jodhpur, many nadis are severely polluted because of poor maintenance and negligence, destruction of catchment areas, and uncontrolled urbanization (Mishra 2001).

Artificial recharge operations may affect groundwater-dependent ecosystems via changes in water table elevation and as a result of water quality changes. People often take it for granted that if water looks clean or a little turbid, and there is no smell and bad taste, then it is of good quality. However, water may be polluted with invisible chemicals, and changes to water availability and quality may be affecting the users. It is also imperative to assess whether farmland has good recharge potential as these areas could accommodate deep percolation with little risk of crop damage or groundwater contamination. If native groundwater is saline, then elevating water tables even seasonally by artificial recharge operations would have the impact of transporting salt into the root zone, with adverse consequences for salt-intolerant species. Therefore, it is essential to ascertain whether on-farm flooding of fertilized land or saline soil will allow the chemicals to leach into the groundwater, which could decrease groundwater quality and also have adverse effects on ecosystems.

Rising water tables can result in anoxia within the root zone, which will stress plants. If the rate of water table fall or rise exceeds that to which groundwater fauna and flora can adapt then ecosystem health will be affected. Changes to the aquifer redox environment are clearly the main potential water quality impact of artificial recharge (AR) schemes. The redox environment can be modified by adding significant amounts of reductants (particulate organic carbon, dissolved organic carbon, NH_4^+, etc.) or oxidants (O_2, NO_3^-, etc.) into aquifers. Adding reductants increases the risk of generating anoxic conditions in the subsurface, while adding oxidants to "reduced" aquifers can increase the weathering rates of some minerals (pyrite, etc.), with potential consequences for increased release of heavy metals, etc. Such data would give a more confident basis for regulators and proponents of AR projects to design, assess, and manage projects to ensure environmental protection.

6.2.7 Basin-Level Impacts

In India, the states of Gujarat, Rajasthan, Maharashtra, Tamil Nadu, Karnataka, Andhra Pradesh, Madhya Pradesh, Orissa and Chhattisgarh have taken up programs on rainwater harvesting and groundwater recharge on a large scale. However, as described in Chapter 3, a major part of these regions is covered by water-scarce river basin systems, which have less than 1000 m^3 of renewable water per annum. Lower rainfall, coupled with higher PET, reduces the runoff potential and promotes high evaporation from the impounded runoff, thereby increasing the dryness. Gujarat and Rajasthan have significant areas of extremely low rainfall. In Maharashtra, Andhra Pradesh, Madhya Pradesh, Karnataka, and Tamil Nadu, over two-thirds of the area falls in the medium rainfall regime. As regards PE, most of Gujarat and Rajasthan and large parts of the other states have medium to high evaporation, and Orissa and Chhattisgarh have medium evaporation. During the monsoon, significant parts of these states experience rain regimes of low to medium to high number of rainy days, low to medium to high rainfall variability, which coincides with medium rainfall/medium to high evaporation, low rainfall/very high evaporation, and high rainfall/medium evaporation.

The spatial variability of rainfall, PE, and number of rain days suggest that the interannual variability in rainfall increases with reducing rainfall; the number of wet spells decreases and the PE increases with decreasing amount of rainfall. The implications of this trend on the potential for water harvesting in low rainfall regions would be: (i) the runoff potential by and large would be low due to a high dryness ratio; (ii) evaporation from surface storage would be high due to high PET; and (iii) the probability of occurrence would be low. As described in Chapter 3, in these states, the percentage of rainfall that recharges groundwater is also quite low, suggesting that the scope for rainwater harvesting to recharge groundwater is very limited in these states. The volume of runoff that would be available for replenishment through natural and artificial recharge from within the local village will therefore be a small fraction of the total consumptive use. Hence, even if the entire runoff generated is harnessed for recharge, it would amount to only a very small percentage of the overdraft. This means that the area has to depend on exogenous sources of water to keep water use sustainable.

Although the impact of upstream RWH and artificial recharge on downstream users may differ from basin to basin, all water uses are likely to affect users elsewhere in the system. This is greatest when all available supplies in the basins have been fully allocated, and in cases with marked inter- and intra-annual variation in rainfall, resulting in decline of runoff coefficient to almost half of its original value, especially when the flow to the reservoirs began to decline sharply due to the construction of thousands of check dams and percolation ponds within the watershed. Indirect recharge and retrieval methods also had a negative impact in managing large irrigation systems due to siphoning of considerable quantities of water to areas not included in the command area plan, and the water recovered consists of a small fraction of groundwater while a larger fraction of abstracted water comes from rivers or lakes. These effects suggest a need for holistic action for water management.

Although modeling tools at regional and spatial scales can help in evaluating MAR scenarios and screening potential sites, testing operational scenarios and hydrological conditions, integrated with other management options, yet identifying areas suitable for MAR and estimating the influence of MAR on a regional basis can be difficult, as there may be limited data on complex surface and subsurface conditions, groundwater levels and flows. Some of these questions can be resolved through field testing, but small-scale pilot field studies can be expensive and may provide limited spatial information. Studies based on GIS modeling in part of the Pajaro Valley Groundwater Basin, California, suggest that only about 7% (15 km^2) of the basin may be suitable for MAR. In addition, there is a need to determine how the impacts of MAR could vary with project location, size, and operating conditions. Modeling suggests that simulated MAR projects placed near the coast can partially help to reduce sea water intrusion but these projects also result in increase in groundwater flows to the ocean. In contrast, projects placed farther inland result in more long-term reduction in sea water intrusion and less groundwater flowing to the ocean.

Further research is recommended to strengthen and extend these findings. It is reasonable to assume that there is an effect of seasonality present in the data collected during this study. In order to improve this analysis, consistent data collected over multiple years are necessary. Since the use of tracers shows great potential for future groundwater studies related to rainwater harvesting, it is recommended that future research extends the length of the study period, utilizes additional tracers to quantify rates of recharge and age of groundwater, for example $^3H/^3He$, and applies these tools to different sites in order to build a reliable dataset to guide future water management planning in the basin.

6.3 Issues/Limitations/Questions that Need Attention

Rainwater harvesting and related water management interventions are mostly based on a deep-rooted belief that the greater the size of the water-impounding structure, the higher will be the benefit in terms of water storage and recharge. Very often, such belief is based on limited knowledge of the local hydrological regimes, which govern the potential supplies of water for harvesting. The reasons for this are the lack of data on stream flows for small rainwater catchments, and ignoring the fact that runoff harvesting is most suited to areas with high runoff catchment area to run-on area ratio. The higher the aridity, the larger would be the required catchment area for the same water yield.

Managed groundwater recharge plans (MGRP), which generally include numerous percolation tanks, check dams, nala bunds, anicuts, weirs, cement plugs, recharge shafts/dug well structures, and revival of ponds, are quite expensive and usually make no mention of the private and social benefits that such investment would generate. In India, approaches including Managing Aquifer Recharge and Sustainable Groundwater

Use through Village-level Intervention (MARVI), Andhra Pradesh Farmer Managed Groundwater Systems (APFAMGS), Social Regulations in Water Management (SRWM) in Andhra Pradesh, Pani Panchayat in Maharashtra and Tarun Bharat Sangh (TBS) in Rajasthan differ in a number of ways from each other. Although they are reported to be successful at local level, they have faced challenges with respect to the need for mechanisms and funding sources to sustain them over the long term.

In the MARVI project, the focus was to create trained village champions/leaders for groundwater monitoring and prepare them to act as an interface between farmers and support agencies, to help facilitate change in farmer behavior and supply them with local data for groundwater understanding and improvement. In the APFAMGS, it was assumed that access to scientific data and groundwater knowledge alone would help farmers make better decisions regarding groundwater use, and so the use of groundwater by farmers in the area could not be regulated. Other approaches such as SRWM and TBS were successful to a limited extent in the areas where they were tried but could not spread to other parts of India due to lack of inputs required for sustainable groundwater management.

Rainwater harvesting does not simply involve constructing recharge structures; it is necessary to repair and maintain these structures over time, and bring them into readiness for groundwater recharge before the monsoon. Such high-cost plans can be defended only if a well-thought-out road map is created to ensure the future sustainability of structures. In many river basins, the surface water systems and groundwater systems are often interconnected. Any alterations made in one system could change the availability of water in the other. In many hilly areas, water levels rise steeply after the monsoon, and groundwater contributes significantly to the stream flows downstream during lean seasons due to the steep gradients for groundwater flow. In such cases, any water harvesting intervention to store water underground may not make much sense as the water stored would be rejected and appear as surface flows. On the other hand, in regions with deep water table conditions, the runoff directly moves into the groundwater systems of the plains through sandy river beds as dewatering of the upper aquifers increases the rate and quantity of percolation.

Although, in some developing and developed countries, large dams have attracted heavy criticism, small-scale "traditional" RWHS as an alternative cannot be widely accepted due to inadequate hydrological and hydrogeological knowledge and understanding. Issues include the source water quality in relation to environmental values of the aquifer, high salinity, intended uses of recovered water, potential for clogging and mineral reactions, and whether groundwater needs to be used for drinking supplies or high conservation ecological values. Also whether the aquifer is confined or artesian, fractured or cavernous, if there are similar projects with similar source water in the same aquifer and whether the project is likely to attract attention under local planning or development regulations.

While harnessing water for recharge is extremely important during normal and wet years, the natural recharge in hard-rock formations is high during such years, depending on seasonal rainfall, further reducing the scope for artificial recharge. Recharge structures remain in use generally for 3–5 monsoon rainfall months and their deterioration tends to be rapid when in disuse. Decades of experience of watershed management programs in India prove that generally conditions of most of the government and common property structures (except dug wells) rapidly decline and become unusable because of the absence of community involvement in their maintenance and repair. Even rehabilitated tanks soon require another rehabilitation program and modernized canal irrigation systems decline in a few years for lack of maintenance.

Unfortunately, in the planning of small rainwater harvesting systems, lack of concern and emphasis on economics and cost seem to remain partly due to the lack of scientific understanding of the hydrological aspects of small-scale interventions, such as the volume of stream flow available at the point of impoundment, its pattern, the amount that could be impounded or recharged, and the influence area of the recharge system. Moreover, there are great difficulties in generating vital data at the micro level on daily rainfall, soil infiltration rates, catchment slopes, land cover and PET, which determine the potential inflows, and evaporation rates, which determine the potential outflows. Further, for small water harvesting projects implemented by local agencies and NGOs with small budgets, it becomes difficult to justify the cost of hydrological investigations and planning. Often, provision for such items is not made in small projects.

In the Indian context, with two-thirds of the geographical area underlain by hard-rock formations, the storage capacity of aquifers poses a major challenge for artificial recharge. Most parts of water-scarce states are underlain by hard rocks such as basalt, crystalline granite, and sandstone. A small area in Gujarat has extensive alluvium: the Narmada Valley and Cambay Basin. The hard-rock aquifers have no primary porosity and have only secondary porosity. The constraints imposed by hard-rock geology in recharge efforts through percolation tanks include high depth to the water table below and around the recharge structure due to the occurrence of recharge mounts and shallow bed rocks, which prevent the percolation of water; and low infiltration capacity of the thin soils overlaying the hard-rock formations. Due to low specific yield (0.01–0.03), the sharp rise in water levels is observed in aquifers during the monsoon, leaving little space for infiltration from structures.

In India, hardly one-fourth of the total area of the aquifers has been mapped by central government so far, which is not sufficient to make a visible impact on the ground for aquifer management (including their recharging) along with stakeholder participation. To effectively implement such programs, groundwater management at the local level does not exist. With multiple owning of wells and tube wells and overlapping jurisdictions of common groundwater resources, the legislative and policy response to address this problem has yet to take place. As described earlier, both "traditional" and "modern" WHSs have strengths and weaknesses. Therefore, even if the existing "WHS" structures are not really "dying," the proponents of "revival needs" are reinterpreting this to make it look like "dying," to suit their organizations' wish for survival and support. Wisdom (ancient or modern) never dies.

The historical phases background described in Chapter 4 clearly suggests that more emphasis on local control on natural resources management and significant internatonal funding support to NGOs for community-based watershed management has occurred between the late 1980s and today. Both the public and the government became more aware of the significance of WHS, due to campaigns by NGOs and the media, declining levels of groundwater and water scarcity in many parts. Despite a lot of publicity about water harvesting systems (WHS) technologies over recent decades, due to failure to arrest groundwater decline, lack of maintenance or declining conditions of a number of WHS, the need to revive the traditional WHS has dominated the development discourse.

It has to be kept in mind that the so-called "traditional" was not necessarily always benign and was small-scale through privately owned wells; in ancient times, secured irrigation facilities were also extended through medium- to large-size bunds and reservoirs. Moreover, since rainwater and groundwater are two interdependent water sources, farmers who use the alternative water source create an additional negative impact on groundwater users, by diverting part of the recharge of the aquifer.

The choice regarding groundwater extraction and rainwater harvesting depends both on time and on the current state of the resource. Local communities may not always like to have a WHS if it does not lead to better access to water or increased benefits in terms of harvest or employment. Farmers may have various long-term irrigation strategies: irrigation relying only on groundwater, on rainwater, or on a combination of both. The use of these strategies depends on both costs and qualitative considerations. It would be unethical as well as unfair to turn a blind eye to the indiscriminate water extraction and focus exclusively on water conservation.

Another issue that has to be examined is the ownership of water entitlements, such as access to parts of a lake or river. The owner can then sell the entitlement to someone else or give another access to the water for a fee. As water resource issues are influencing projects and policies in many government departments, it is essential that responsible agencies work together to further develop relevant schemes. Investment opportunities will be created for reviving traditional technology. Studies on water balances and the interaction between surface and subsurface reservoirs, a critical issue, would also help greatly in understanding the value of the practice.

It is hoped that with further information from SWOT analyses of the operation of such small-scale RWH schemes, opportunities for meaningful large-scale RWH schemes will be developed to further reduce the pressure on limited water resources, and its unplanned overabstraction and use. This would require widespread dissemination of information on the good work done so far by water resources professionals, so that aspects such as utilities, resource assessment, and societal use can be discussed and financed by society as a whole.

6.3.1 Managed Aquifer Recharge

For meaningful application of MAR, the presence of a suitable aquifer, which can store and convey large volumes of water, is an essential prerequisite, because the unit cost of recovered water reduces with increase in storage volume. Therefore, to maximize water recovery ability, thick consolidated aquifers having uniform hydraulic properties with a very low regional water flow rate through the aquifer are preferred, compared to unconsolidated ones, due to simpler well construction and ease of maintenance. However, for MAR, there can be both positive and negative aspects of other aquifer attributes. For example, in unconfined aquifers, infiltration methods may be used, which are cheaper than well injection methods if sufficient land is available. However, stored water needs to be protected from pollution, especially where recovery is for drinking water supplies. Confined systems are better protected from pollution but can be accessed only by wells. For recharge wells, the water quality requirements to avoid clogging are more stringent than for surface infiltration systems, and depend on pore sizes in the aquifer, its mineral composition, and the form of construction of the well.

If the ambient groundwater is brackish, the water pretreatment requirements for aquifer protection may be less. However, if the groundwater is too saline, recovery efficiency may be low and the site is non-viable. Reactive minerals in aquifers, such as carbonate, can assist in controlling clogging but the same minerals can in some cases also contain metals that are released and impair the quality of recovered water. The oxygen content of the aquifer can also affect water quality. Pathogens and some organics are most effectively removed under aerobic conditions but other organics are only removed under anoxic conditions. To get the best water quality improvement, it is ideal to have different zones in the aquifer strata, separated by impervious or low-permeability layers, so that water is exposed to different conditions. This allows choice of the most favorable characteristics for water storage. At any specific location, depending on the degree of interconnection of different aquifer strata, it may be possible to store water of different qualities in different aquifers.

In many locations, MAR is not feasible if there is no aquifer with suitable characteristics to allow sufficient storage while ensuring environmental protection; for example, the aquifer is unconfined and the water table is very shallow; the aquifer is very thin or composed of fine-grained unconsolidated material; the site is adjacent to a leaky fault or a semi-confining layer containing poor-quality water; or the aquifer contains poor-quality water and is highly heterogeneous or has a high lateral flow rate. Hence, local hydrogeological knowledge is needed to identify the presence of aquifers and suitability for MAR. State departments responsible for groundwater generally require drillers to lodge basic stratigraphic and hydrogeological information for each well drilled and this information is stored in departmental databases. Hydrogeological reports generally provide some indication of the level of knowledge about local aquifers and their degree of uniformity. However, since aquifer properties vary spatially, when assessing viability of MAR it is inadvisable to extrapolate data from one site to a nearby site.

At basin level, in both rural and urban areas, in places where groundwater levels have been in decline, MAR alone may not be sufficient to restore groundwater equilibrium. Appropriate resource management to prevent excessive use of groundwater may also be needed. Due to the large numbers of well owners, new governance methods may be required involving collective management. Costs imposed by restricting groundwater use may be compared with costs of MAR to determine optimal strategies in relation to changes in climate, land use, and the value of various uses of water. MAR should be avoided in overallocated surface water catchments, as its use would further deplete environmental flows and availability of water to meet downstream water entitlements.

However, in many other parts of the world, including India, in general, farmers' attitudes to recharge tend to be opportunistic, and compared to the best available recharge sites, they prefer sites that can benefit their own wells. Moreover, in reserve forestlands, only government departments are allowed to construct and operate recharge structures. Also, farming communities cannot choose small recharge structures consistent with their resources and skills. Therefore, in Gujarat, despite government interest, support and incentives, farmers in many groundwater-depleted districts show little enthusiasm for recharge. For example, in the north Gujarat alluvial aquifer area, despite quite severe groundwater depletion, farmers are lukewarm about decentralized recharge work, interest remains very limited, and energy use in pumping is very high. In these areas with porous and permeable aquifers, recharge results in flat recharge mounds with little impact on water levels in wells.

In most cases, the basic MAR strategy pays little attention to whether the areas identified for MAR are in real need, and takes a narrow view of the costs and benefits of alternative action, and runoff allocation. However, in reality, if reallocating a certain amount of runoff to MAR benefits the local society more, compared to diverting it to past investments in surface storages, MAR allocation should be the chosen course of action, even if surface storage is the principal source of irrigation and domestic water supply. This demand-side perspective needs to be included when evolving a MAR strategy.

For instance, based on data compiled by Gujarat government agencies, groundwater is only 30% of Gujarat's total water resources, but it meets 82% of the state's domestic water needs, 65% of industrial water demand, and serves 79% of the state's gross irrigated area. Gujarat needs over 13 500 m^3 of surface storage to irrigate a 1 ha area. While part of this contributes to soil moisture and groundwater recharge, a significant amount is lost to evaporation. In contrast, irrigating 1 ha with tube wells uses only 3500 m^3 of groundwater. A million m^3 removed from surface storage and dispatched to aquifers reduces the canal-irrigated area by 72.2 ha but increases groundwater irrigation by 285 ha. Even committing 80% of surplus runoff generated within each basin is insufficient to cover the accumulated groundwater deficit in Kutch and Saurashtra in less than a century.

6.4 Opportunities for Revival of WHS

Groundwater recharge with naturally treated storm water or reclaimed water has been practiced in the USA since the 1960s for recovery for non-potable and drinking water supplies, and is well accepted by the public for potable purposes (Asano and Cotruvo 2004). In any city, town or rural area, the costs and benefits of MAR can be compared with a range of options including improved water conservation, tapping new surface water supplies or aquifers, rainwater tanks, and groundwater or sea water desalination. The costs of each option depend on the local situation, and large variations occur between localities for costs of any option and therefore the relative cost effectiveness of different options. The unit cost of infiltration ponds and soil aquifer treatment practiced in rural settings depends strongly on the infiltration rate in the basins, and rates vary between basins and depend on permeability of the soil and depth of the groundwater table. For example, in a pit with half the infiltration rate, the cost per kilolitre may be approximately double.

Decision makers should be provided with integrated information regarding WHS impacts and policy options for water scarcity mitigation and adaptation at states and national level, and about their respective costs, benefits, and risks to enable them to assess the consequences associated with alternative choices. More research needs to be done to enhance knowledge for efficient diagnosis in complex environmental matrices to allow rational approaches to identification of source water pretreatment needs and acceptable uses of recovered water. Governments and researchers need to execute their roles at the science–policy interface in carefully recognizing and managing diverse viewpoints, without compromise on scientific rigor and objectivity. It would be better to avoid multiple geographical operating team and technical support units so that management tasks are not further complicated, and the best researchers' participation with high time and travel commitments is not put at risk. For effectiveness, baseline work needs to be done by scientifically independent organizations, which are not politically or economically driven.

The storage and treatment values may vary from aquifer to aquifer, but design of recharge operations in accordance with prescribed guidelines should ensure that the values are achieved. Aquifers also have a tangible value for treatment, with the advantage that this treatment is passive, requiring no further input of energy or chemicals. For instance, the extensive fresh-water aquifers have significant storage value, especially in already established urban areas where land value is high, and the transmission value is similar to the treatment value and possibly of higher value. But, in saline aquifers, long-term storage value may be diminished by a higher depreciation rate, depending on the mixing of fresh injected water with native saline groundwater. In brackish or localized aquifers, the transmission value is negligible. This restricts the proportion of injected water that can be recovered at an acceptable quality.

Overall knowledge of the existence and properties of aquifers has potentially huge value for urban areas in allowing more cost-efficient future water supply options, particularly for interseasonal storage. Investment in hydrogeological investigations may reveal latent water supply infrastructure, and in some cases may identify viable subsurface storage alternatives to unacceptable or expensive projects. For successful implementation of MAR at aquifer or basin level, the plan needs to take into consideration the hydrogeological principles, water demand and supply, storage space and runoff, rainfall–runoff relationship, soil moisture movement, spatial and temporal variability of water flows and fluxes, surface–groundwater interactions, and consumptive and non-consumptive water uses. In general, the usefulness of MAR depends on the following aspects.

6.4.1 Demand

Before adopting MAR, it is essential to define the purposes and estimate the demand for recovered water (within an economic scale) or a clearly defined benefit of recharge. In urban areas, demand for storm-water detention to mitigate floods, improving coastal water quality and enhancing urban amenity and land value may also contribute revenue for MAR projects. For reclaimed water projects, the decline in discharge of treated effluent to sea may provide a motivation for investment in MAR.

6.4.2 Source

Since, in an overallocated catchment, it is quite likely that an entitlement to surface water would not be available, entitlement to water to be used for recharge needs to be secured. The average annual recharge quantity should be at least commensurate with mean annual demand, and should be in sufficient excess to build up a buffer storage to meet reliability and quality requirements.

6.4.3 Aquifer

A suitable aquifer for MAR needs to have an adequate rate of recharge and sufficient storage capacity, and be capable of retaining the water where it can be recovered.

Low-salinity and marginally brackish aquifers are preferable so that mixing with fresh recharge water should allow recovered water to be fit for use.

6.4.4 Detention Storage

There should be open space, or dams, wetlands, ponds or basins to detain sufficient water without causing flood damage to enable the target volume of recharge to be achieved. In established urban areas, space can be a major impediment to storm-water harvesting, and to avoid land requirements of infiltration systems, recharge wells are commonly used. For recycled water from a sewage treatment plant, generally no additional detention storage will be required at the recharge facility.

6.4.5 Management Capability

To meet governance requirements, it is essential to have detailed investigations and design of projects, involving experienced expertise with hydrogeological and geotechnical knowledge, as well as knowledge of water storage and treatment design, water quality management, water-sensitive urban design, hydrology and modeling, monitoring, and reporting.

6.4.6 Water Quality Hazards

The aquifer response to any hazard depends on specific conditions within the aquifer, including temperature, presence of oxygen, nitrate, fluoride, organic carbon and other nutrients and minerals, and prior exposure to the hazard. To assess risks and determine sizes of attenuation zones and siting of monitoring wells, the attenuation rates of pathogens and organic compounds under a range of conditions should be considered. In most aquifers, and with appropriate pretreatment of water to be recharged, the attenuation zone is small, generally of the order of 20–200 m from the recharge area or well. Water that travels laterally has sufficient residence time in the aquifer for attenuation of pathogens and contaminants to below the relevant guideline values for native groundwater and intended uses of recovered water.

The zone of aquifer in which water quality may be affected by MAR may be larger, but in this outer domain the water quality should continuously satisfy the initial environmental values of the aquifer. The effects of MAR operations on hydraulic heads (pressures) may be measurable over a much larger area, especially in confined aquifers. If the aquifer is originally too saline for the uses of recovered water, a storage zone can be identified that contains water which, when recovered, is fit for its intended use. The outer boundary of the attenuation zone represents the maximum separation distance between the MAR recharge structure and well(s) for verification monitoring to ensure that the ambient groundwater quality is protected. Both during operation of MAR and on cessation of the MAR operation, the boundaries of the various zones do not remain sharp (as shown in Figure 6.1), and may shrink and disappear, and remain diffused.

6.4.7 Role of Ethics

In order to achieve really meaningful results from the WHS, the environmental, economic, social, geographical, and political aspects should be taken into account, and

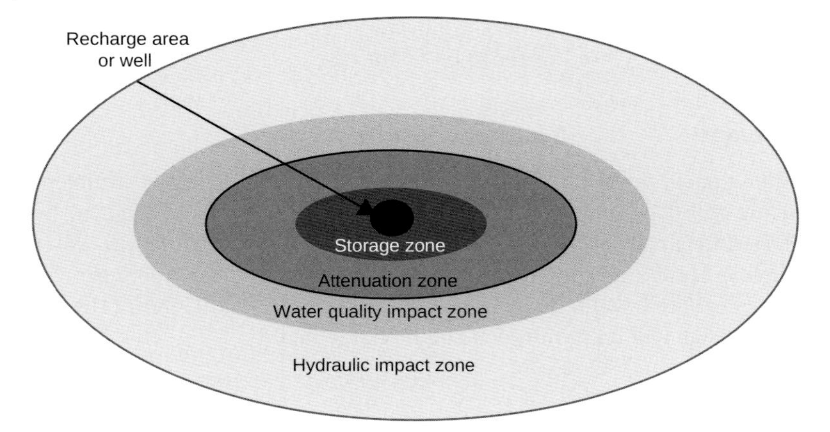

Figure 6.1 Schematic showing zones of influence of a recharge operation. Recharge area or well; storage zone; attenuation zone (limit shown by solid line); water quality impact zone; hydraulic impact zone.

ethically govern the social actions and practical decisions of the individual/group with moral values (honesty, trust, and treating others fairly). In the context of the WHS, "geo-ethics" gives an opportunity for the geo-sciences to raise public awareness and knowledge, and help to understand the great complementarity and interconnection between natural systems (rocks, soils, sediments, minerals, water, landforms, fossils, flora and fauna and the processes, such as sediment transport, that shape them throughout geological time) and human systems (link between people, their culture, and the natural landscape). The time has come to educate all stakeholders (including politicians, policy makers, media, and public) to consider "geo-ethics" in their choices, from the best obtainable scientific information and socioeconomic considerations (Datta 2013, 2015). This would help to manage, prevent, mitigate, adapt, and overcome impacts of unexpected hazardous consequences, by critical analysis of ethical dilemmas (Datta 2013, 2015).

To ensure selection of the proper location of sites for construction and maintenance of infrastructures, and to characterize spatiotemporal variability of geo-regimes, integrated interdisciplinary approaches by reliable, competent, and transparent institutions can be helpful. Involvement of corrupt and unscrupulous practices for private gain should be discouraged, to enhance efficiency and promote harmony through trust, knowledge, transparency, respect of interests, and share of responsibility/resources.

6.4.8 Role of Geoscientists

In any watershed system, when surplus runoff occurs, this causes morphing of the usual catchments and streams, totally altering the landscape, channels, and stream heights. Generally, local knowledge of the geological system and landform heterogeneity is limited and insufficient for surplus runoff management response requirements (Datta 2013, 2015). Regular remapping of watershed catchments is desirable. In order to manage, prevent, mitigate, and harvest high-intensity rainfall-induced floods, geoscientists with wider experience can play a significant role in mapping and analyzing changes in topography, geology, land contours, stream and river flows, flow paths, river beds and flood plains, for clear land classification and hydrological calculations, and to ensure proper selection of project location, planning, preliminary engineering design, construction, infrastructure maintenance, and preparedness strategies.

These scientists can also involve school children in programs at the local level to educate and create awareness in the local population (including politicians, administrators, policy makers, media, etc.) through technical and scientific knowledge advancements, new communication strategies, and by interaction with politicians, media, and citizens (Datta 2013, 2015). This can be helpful in promoting ethical, cultural, socioeconomic, and educational values, and changing mindsets, strengthening relationships, framing policy guidelines on water conservation, and developing a problem-solving holistic interdisciplinary approach, planned in association with the public (Datta 2013, 2015). They can also educate the public about groundwater resource availability (in terms of quantity and quality) and significance of its proper use, and explain the benefits of its safe extraction, geo- and hydrosphere contamination, to protect groundwater from depletion and pollution, etc. to generate more efficient methods in floodwater management and conservation.

6.5 Threats to WHS in India

Deteriorating conditions and improper maintenance of WHS result in the occurrence of unexpected vulnerable situations, mostly due to lack of harmony between natural and human systems, failure to introduce proper preparedness measures, and tendency to manage vulnerability by *ad hoc* approaches with no ethical framework (Datta 2013, 2015). Current priorities for WH seem to be guided by providing immediate solutions to present problems. Absence of coherent long-term policies on communication by the mass media with half-baked information and generic opinions from all kinds of opinion makers also creates public panic due to regional polarization, with self-interest to the fore.

The WHSs are also under threat due to unclear understanding of the hidden complexities in power relations and agendas of the different actors within the "local" and "traditional" camps. This is because the boundaries of the "traditional" and "modern" are very vaguely understood. Both investors and those seeking funding support largely ignore such issues, as their main interest remains focusing on the discursive dimension on development. Further, it is desirable to investigate the practice of making money from the water crisis by owning shares in companies involved in improving water quality, construction of desalination plants, infrastructure operations working on dams and water treatment units in developing countries.

Planning for groundwater recharge in India needs to make a distinction between alluvial aquifer areas of the Indo-Gangetic plains and hard-rock aquifer areas of inland peninsular India. In the water-abundant Ganga basin, with high water tables and perennial postmonsoon flooding, large-scale recharge projects are neither needed nor demanded by the public. But, regional allocation of funds for WHS suggests a clear disconnect between groundwater recharge management plans (GRMP) and stakeholder objectives. For example, farmers in Bihar, West Bengal, Assam, and eastern Uttar Pradesh will have limited interest in a groundwater recharge program because they have no scarcity of water, but regions where groundwater depletion is a great concern receive no special attention from the GRMP.

The GRMP also fails to recognize that farmers in alluvial aquifer areas are lukewarm towards recharge projects because they do not directly benefit from recharge mounds that rapidly dissipate. In contrast, farmers in many hard-rock areas of India have taken to recharge programs with great enthusiasm. While the villagers want recharge structures like check dams sited near farming areas that demonstrably can benefit their existing wells and new wells to be dug, the GRMP emphasizes siting of recharge structures to maximize recharge no matter where. For instance, in arid alluvial areas of north-western India, limited rainfall offers little scope for rainwater harvesting for natural recharge but in hard-rock aquifer areas, there is need for a special program of groundwater recharge.

References

Asano, T. and Cotruvo, J.A. (2004). Groundwater recharge with reclaimed municipal wastewater: health and regulatory considerations. Water Research 38 (8): 1941–1951.

Borthakur, S. (2008). Traditional rain water harvesting techniques and its applicability. Indian Journal of Traditional Knowledge 8 (4): 525–530.

Central Ground Water Board (CGWB) (1995). Report on Groundwater Resources of India. New Delhi: Government of India.

Chawla, A.S. (2000). Ground Water Recharge Studies in Madhya Ganga Canal Project. Colombo: Consultancy Report to IWMI.

Das, S. (2010). Johads of Alwar. Journal of Geological Society of India 75 (2): 446–447.

Datta, P.S. (2013). Groundwater vulnerability to changes in land use and society in India. Understanding Fresh-water Quality Problems in a Changing World. Proceedings of H04, IAHS-IAPSO-IASPEI Assembly, July, Gothenburg, pp. 345–352.

Datta P.S. (2015). Traditional Rainwater Harvesting at Crossroads – Time to Seek Ethical Opportunities for Reform. Proceedings of the Aqua Foundation IX World Aqua International Congress, Reviving Traditional Water & Environment Conservation Techniques, November 26–27, New Delhi, pp. 93–115.

Datta, P.S., Desai, B.I., and Gupta, S.K. (1980). Hydrological investigations in Sabarmati basin-I. groundwater recharge estimation using tritium tagging method. Proceedings of the Indian Academy of Science Physical Science 46 (1): 84–98.

Datta, P.S. and Goel, P.S. (1977). Groundwater recharge in Punjab state (India) using tritium tracer. Nordic Hydrology 8: 225–236.

Datta, P.S., Goel, P.S., and Rama Sangal, S.P. (1973). Groundwater recharge in western Uttar Pradesh. Proceedings of the Indian Academy of Science 78 (Sec. A): 1–12.

Mishra, A. (2001). The Radiant Raindrops of Rajasthan. Dehradun: Research Foundation for Science, Technology and Ecology.

Muralidharan, D. and Athavale, R.N. (1998). Base Paper on Artificial Recharge in India. Hyderabad: National Geophysical Research Institute, CSRI.

Raju, K.C.B. (1998). Importance of recharging depleted aquifers: state of the art of artificial recharge in India. Journal of Geological Society of India 51: 429–454.

Sakthivadivel, R. and Gomathinayagam, P. (2004) Case studies of locally managed tank systems in Karnataka, Andhra Pradesh, Gujarat, Madhya Pradesh, Gujarat, Orissa, and Maharashtra. Report submitted to IWMI-TATA Policy Programme, Anand.

Sharma, A. (2006). Water harvesting context in the Indian Subcontinent. UNESCO G-WADI Meeting on Water Harvesting, Aleppo, November 20–22, pp. 63–70.

Shingi, P.M. and Asopa, V.N. (2002). Independent Evaluation of Check Dams in Gujarat: Strategies and Impacts. Executive Summary. Ahmedabad: Centre for Management in Agriculture, Indian Institute of Management.

Swedish International Development Cooperation Agency (SIDA). (2013) SIDA Decentralised Evaluation 2013:28. Available at: www.sida.se/publications.

Further Reading

Amos, C.C., Rahman, A., and Gathenya, J.M. (2016). Economic analysis and feasibility of rainwater harvesting systems in urban and peri-urban environments: a review of the global situation with a special focus on Australia and Kenya. Water 2016 (8): 149.

Datta, P.S. (2014). Need for Better Groundwater Governance for Sustained Water Supply. In: IPHE National Conference on Piped Water Supply and Sewerage Systems, January 24–25. New: Delhi.

Gupta, S. (2011). Demystifying "tradition": the politics of rainwater harvesting in rural Rajasthan, India. Water Alternatives 4 (3): 347–364.

Imteaz, M.A., Ahsan, A., Naser, J., and Rahman, A. (2011). Reliability analysis of rainwater tanks in Melbourne using daily water balance model. Resources, Conservation and Recycling 56: 80–86.

Molden, D. and Sakthivadivel, R. (1999). Water accounting to assess use and productivity of water. International Journal of Water Resources Development 15 (1 & 2): 55–71.

Mudgal, K.D., Kumari, M., and Sharma, D.K. (2009). Hydrochemical analysis of drinking water quality of Alwar District, Rajasthan. Nature and Science 7 (2): 30–39.

Neelakantan, S. (2003) A Gossipmonger's Revisit to Chettipalayam. Working Paper No. 142. Madras Institute of Development Studies, Chennai.

Parthasarathi, G.S. and Patel, A.S. (1997). Groundwater Recharge Through People's Participation in Jamnagar Region, 51–56. Nagpur: Indian Water Works Association.

Rahman, A., Dbais, J., and Imteaz, M. (2010). Sustainability of rainwater harvesting systems in multistory residential buildings. American Journal of Engineering and Applied Sciences 3 (1): 889–898.

Rahman, A., Keane, J., and Imteaz, M.A. (2012). Rainwater harvesting in greater Sydney: water savings, reliability and economic benefits. Resources, Conservation and Recycling 61: 16–21.

Roebuck, R., Oltean-Dumbrava, C., and Tait, S. (2011). Whole life cost performance of domestic rainwater harvesting systems in the United Kingdom. Water and Environment Journal 25: 355–365.

Roebuck, R.M., Oltean-Dumbrava, C., and Tait, S. (2012). Can simplified design methods for domestic rainwater harvesting systems produce realistic water-saving and financial predictions? Water and Environment Journal 26: 352–360.

Shah, T. (1998). The Deepening Divide: Diverse Responses to the Challenge of Ground Water Depletion in Gujarat. Policy School, Anand.

Shah, T. (2008). India's master plan for groundwater recharge: an assessment and some suggestions for revision. Economic and Political Weekly 43 (51): 41–49.

Sinha, S., Gururani, S., and Greenberg, B. (1997). The "new traditionalist" discourse of Indian environmentalism. Journal of Peasant Studies 24 (3): 65–99.

Stiefel, J.M., Melesse, A.M., McClain, M.E. et al. (2009). Effects of rainwater-harvesting-induced artificial recharge on the groundwater of wells in Rajasthan, India. Hydrogeology Journal 17: 2061–2073.

Sukhija, B.S., Reddy, D.V., Nagabhushanam, P., and Nandakumar, M.V. (2005). Efficacy of percolation ponds as artificial recharge structures and the controlling factors. Journal of the Geological Society of India 66: 95–104.

Yang, S.-Q. (2015). Coastal reservoir – a technology that may dominate future water supply. Journal of Water Resource and Hydraulic Engineering 4 (4): 388–397.

7

Challenges Associated with Water Harvesting

CHAPTER MENU

7.1 Water Bodies and Associated Problems and Challenges

Water bodies are generally discrete and significant surface water such as a lake, a reservoir, a stream, a river or canal, part of a stream, traditional water storage structures or a stretch of coastal water and similar structures, potentially vital in the development of the rural economy in developing countries. Through the ages, such natural and man-made water bodies have sustained agriculture. Also, traditional water bodies are used in rural areas, which, *inter alia*, include various purposes such as domestic, drinking, irrigation, horticulture, etc. Water bodies serve as storage reservoirs in monsoon-dependent areas where there is a shorter period of rainfall and a long dry spell with very high deviation of annual rainfall. Therefore, there is an imperative need to ensure proper, efficient and sustainable management and development of water bodies all over the world through sustained efforts, funds, and programs primarily by governments and by individuals, groups, institutions, and local bodies.

The different types of rainwater harvesting (RWH) water bodies have been described earlier. In different parts of India, traditional RWH systems usually include tanks, bhandaras (Maharashtra), phad irrigation (north-western Maharashtra), khadin and baolis (Rajasthan), vav or vavdi (Gujarat), ahar pynes (Bihar), dung or jumpois (Jalpaiguri district, West Bengal), and zing (Ladakh, J&K). Small storage tanks (mostly community owned) are called ponds or bundhis. Government departments or local bodies generally construct the large storage tanks covering command areas of 20–2000 ha. In India, state governments identify the water bodies and give detailed information in respect of each water body, for coverage under various schemes of the central government ministry.

In many parts of the world, particularly in developing countries, there has been deterioration in water quality in water bodies due to indiscriminate disposal of sewage. In most cases, untreated, treated or partially treated domestic sewage is generally

Water Harvesting for Groundwater Management: Issues, Perspectives, Scope, and Challenges, First Edition. Partha Sarathi Datta.

discharged into nearby water bodies, causing severe sanitary and other water pollution problems. Runoff containing agricultural wastes from agricultural fields and animal farms, which consists of agro-chemicals in the form of fertilizers, organic manure, pesticides, etc., may also ultimately enter into water bodies, causing deterioration of water quality. The agricultural runoff is rich in nutrients like nitrogen, phosphorus, organic matter, and pesticides, which causes nitrification toxicity to aquatic life. Industrial waste water has the greatest potential for polluting water bodies. The waste from industries including sugar factories, dairies, paper and pulp, tanneries, distilleries, and metal-plating, which release substantial quantities of heavy metals and organic matter, causes heavy pollution. Mining activities also lead to gradual damage to the surface as well as groundwater resources.

7.2 Land Encroachments, Land Grabbing, Water Pollution, and Other Issues

In the recent past, across the globe, the water supply-oriented paradigm for water resources management has become a part of contemporary patterns, often in the form of land and water resources appropriation from one community or operation to another. The expressions "land and water grabbing" and "land encroachment" essentially refer to a situation in which powerful people or groups, public or private, are able to take control of or reallocate precious land and therein water resources for their own benefit. Such land and water grabbing, by encroachment on catchments for water harvesting systems, have become rampant in the name of water management in areas of water abundance or scarcity, and are common practice in both developing and developed countries in some parts of the world, for industrial purposes, crop cultivation, extractive mining operations, real estate construction, development of energy, food, etc.

These practices often result from lack of proper governance, influenced by certain powerful individuals and groups, often at the expense of the local communities. Such practices can result in devastating effects on the local villagers who are driven from their villages, to make land available for mega-, macro- or micro-water harvesting structures, privatization of water sources, and for industrial purposes that benefit the few. However, limited reliable evidence and proof-based information are available to identify the areas/zones where water bodies have been impacted by encroachment and degraded severely due to unauthorized activities, and the steps taken or proposed to curb the deteriorating water quality trend.

Such land encroachment is turning the water resource from a freely accessible common good to a private good controlled by those in power. In recent years, various powerful groups, industries, etc., from governments to large national and international companies, have appropriated agricultural lands, particularly in developing countries. This process is characterized by large-scale investments in the name of rural development, which often bring little benefit to the common people, instead favoring the stakeholders of large agribusiness corporations. The most common processes are appropriation, privatization, impoverishment, commercialization, and transformation of a freely available natural resource (land and water) from a common good to an economic good, liberalization and privatization (opening to the market and to private management companies), creating a financial asset that can be exchanged on major global equity markets.

In the Indian context, removal of alleged encroachments on water bodies is the responsibility of the relevant state government/local body. As and when any such issue is brought to the notice of the Ministry of Environment and Forests (the nodal ministry for the National Lake Conservation Plan (NLCP)), it is referred to the relevant state government for necessary action at their end. Under the Constitution, water being a state subject, the state governments do not allow the central ministry to make any regulation for control of any water body. This problem exists in all states and in most towns. Encroachment slowly creeps upon the water bodies since they belong to no one. People come and settle down and in time they claim the title and with the help of unscrupulous local politicians, they manage to sell this off. There has been a steady decline of water bodies, especially in urban areas. But, as far as the law is concerned, this is a very local phenomenon and it needs vigilance from municipalities or small town committees to control its spread.

7.2.1 Status of Water Bodies in India

There has been a decline in the number of water bodies throughout the world, due to poor maintenance of traditional water bodies, resulting in reduced storage capacity and lower efficacy. In the Indian context, according to the Third Minor Irrigation Census (2000–2001) (Source: Report of the Third Census of Minor Irrigation Schemes, 2005), there were 5.56 lakh tanks and storages in India under surface flow and surface lift schemes as minor irrigation sources, creating 6.27 million ha of irrigation potential. Out of 5.56 lakh tanks, about 85% are in use, and the remaining 15% are not in use due to one reason or another. Due to non-use of tanks, 1 million ha of irrigation potential is lost. The main factors responsible for water bodies permanently not in use are salinity, dried-up water bodies, destroyed beyond repair, sea-water intrusion, industrial effluents, etc. State governances are responsible for related management issues on various action programs undertaken in the water sector for ensuring water use efficiency.

Regarding remedial measures taken to prevent and arrest deterioration in the quality of water bodies, the Water Quality Assessment Authority (WQAA) was constituted under the Environment Protection Act, 1986. The functions of the WQAA are to direct agencies (government/local bodies/non-governmental) to standardize method(s) for water quality monitoring and to ensure quality of data generation for utilization thereof, to take measures to ensure proper treatment of waste water with a view to restoring the water quality of the water bodies to meet the designated best uses; to take up research and development activities in the area of water quality management; to promote recycling and also reuse of treated sewage/trade effluent for irrigation in development of agriculture; to draw up action plans for quality improvement in water bodies; and monitor and review/assess implementation of remedial schemes.

The responsibilities of the WQAA also include scheme(s) for imposition of restriction in water abstraction and discharge of treated sewage/trade effluent on land, rivers, and other water bodies with a view to mitigating the crisis of water quality; to maintain minimum discharge for sustenance of aquatic life forms in riverine systems; to utilize self-assimilation capacities on critical river stretches to minimize the cost of effluent treatment; to provide information to pollution control authorities to facilitate allocation of waste load; to review the quality of both surface water and groundwater (except that due to geogenic aspects) and identify "hot spots" for taking necessary actions for improvement in water quality; to interact with the authorities/committees constituted under the provisions of the Act for matters relating to management of water resources; to constitute/set-up state-level water quality review committees (WQRC) to co-ordinate the work to be assigned to such committees; and to deal with environmental issues concerning water resources, which may be referred to it by the central or state governments.

With respect to the measures for repair, renovation, and restoration of water bodies, there is a need for maintenance of existing water bodies in rural areas and also to turn water bodies in urban areas into esthetic, decorative objects. In most states, the smaller water bodies are entrusted to the local panchayats for maintenance and management while the bigger ones are vested with the water resources/irrigation departments. The maintenance and management of the water bodies come within the purview of the state government because water is a state subject. However, rivers, ponds, tanks, irrigation systems, canals, and aquifers, which may be public goods, and fall within the realm of common property, or be state owned, may equally run the risk of being captured by large landowners and elites. For instance, large or upstream landholders may block water from flowing downstream. Countries that have large rural populations are also often agrarian societies, where land and water can be highly contested resources.

In order to evaluate the efficacy of the methods presently employed for upkeep of water bodies and to collate the findings of such efforts, the scheme envisages concurrent evaluation and impact assessment by an independent body, which will look into the management of water bodies. Regarding the standards presently employed in maintenance and management of water bodies in India compared with those employed abroad, the subject needs to be studied on the basis of evaluation of existing schemes. However, due to non-availability of data on the number, size, and status of all water bodies across the country, the central and state authorities cannot make any comprehensive and co-ordinated efforts for their repair, renovation, and restoration. Therefore, governance bodies should compile comprehensive and up-to-date information on the exact number and status of water bodies all over the country and place it in the public domain and on the website of the ministry, to serve as a useful tool for policy makers, planners, bureaucrats, water users and other stakeholders, and initiate integrated efforts for repair, renovation, and restoration of water bodies.

Noting that pollution of surface water bodies like rivers, lakes, ponds, tanks, etc. can affect the quality of groundwater in aquifers hydraulically connected with them due to percolation, the Ministry of Water Resources may in conjunction with the Ministry of Environment and Forests issue appropriate advisories to the state governments to institute an effective mechanism for monitoring the water quality in water bodies, so that pollution is arrested and a lasting solution worked out for the long-term interests of the nation. The WQAA is responsible for directing the agencies (government/local bodies/non-governmental) to standardize method(s) for water quality monitoring and also for ensuring quality of data generation. The WQAA reports should be made available to state governments on an annual basis, highlighting specific cases where deterioration in quality of water has reached alarming levels and calling for prompt remedial action.

Considering their vital importance, in order to address reports of worrisome encroachments on water bodies, the Central Government of India cannot behave like a helpless bystander on the grounds that removal of encroachments is a state matter, since the central Ministry has already identified an independent body/agency to look into the method of water bodies management. In most states, the smaller water bodies are entrusted to the local panchayats for maintenance and management while the bigger ones are vested with the water resources/irrigation departments of the states. The efficacy of methods presently employed by local communities all over India for the maintenance and management of water bodies needs to be improved. The independent body may be asked to expeditiously complete the evaluation of the methods presently applied for management of water bodies, rectify the shortcomings, and implement necessary remedial measures, such as imposition of deterrents or penalty provisions effective enough to tackle such encroachments.

7.3 Risks Associated with Managed Aquifer Recharge

In many countries, including India, groundwater depletion and erratic rainfall force local communities and governments to implement water harvesting and recharge structures with the expectation of increasing water availability. From ancient times, harvesting of surplus surface runoff on natural or man-made catchment areas has been used for relief of local water availability. While some efforts showed promising results, most provided only limited water availability to meet local needs. The general public are not aware that the harvestable runoff quantity depends on the catchment characteristics, runoff amount, rainfall frequency and intensity, water infiltration, percolation and recharge rates, and that runoff water collected in ponds and to recharge shallow groundwater is mostly of poor quality. Moreover, enhancement of artificial recharge may work for shallow groundwater but for deeper aquifers, sophisticated injection methods and high-quality alternative water sources are required.

In view of the issues described above, the question remains – "To what extent can rainwater harvesting for managed aquifer recharge help in arresting groundwater table decline?" This is because information on the potential of numerous local efforts on basin-scale water availability, distribution, techno-economic feasibility of various approaches, and cost–benefits from such operations is limited, incomplete, and not well understood. Moreover, due to the heterogeneity and distributed nature of aquifers and their interconnectivity across space and with surface water sources, recharge by one group or community may affect water availability for other neighboring or downstream groups. Some international funding agencies have stopped support for such activities, because impact assessments are qualitative analyses based on mostly secondary information, and limited scientific analysis has been done to actually quantify the rise in water table or the time of flow increase in rivulets (Datta 2014, 2015).

Since there is no way to evaluate the maximum exploitable groundwater volume in advance or in absolute terms, in order to arrest groundwater decline, two broad types of management are usually adopted: (i) approaches such as power pricing, subsidies for efficient technologies, economic policies that discourage water-intensive crops, etc., and (ii) approaches dealing with specific aquifers on the basis of command and control management through a resource regulator. Whichever approach is adopted, the development and management of these resources must be based on adequate knowledge of a clear aggregate situation of groundwater.

But there is much to be done. For instance, in Panama, Guatemala and many other regions, there are 35–51 basins, with numerous sub-basins and micro-basins; organized water committees are limited only to a few basins and groundwater resources have hardly been studied. Uganda, Mozambique, Niger, India, and Pakistan are among countries where the largest numbers of people cannot get clean water within a 30-minute trip. This means millions of people having to take long walks to collect water, which is often dirty and likely to make them ill. In general, water resource is not being sufficiently well managed, due to a lack of appropriate infrastructure, lack of suitable policies and/or their enforcement, resulting in poor service delivery, inefficient use, pollution, bad planning and/or implementation of projects. In many places, there is a lack of both governance and infrastructure.

Surprisingly, despite its obvious importance and value, limited observed data are available on groundwater or on withdrawals and abstractions from aquifers and various other water sources globally. Observed data collection should span at least 30 years at the same location so that natural variability and/or trends can be captured. Supporting such long-term monitoring is often beyond the short-term interests associated with political careers. Therefore, it is necessary to change the mindset of relevant stakeholders and adopt a resilience approach, with a medium- to long-term preventive focus, with the aim of recovering or maintaining good living conditions, rather than isolated actions.

For climate-resilient groundwater management, it may be useful to create an integrated system of adequate water supply, based on spatial and temporal variation in different time scales of groundwater recharge. Since groundwater withdrawals generally have both recent and past recharge components, and the present to past recharge ratio is large in humid areas and small in arid areas, transient and steady-state conditions prevail at low recharge rates and at high recharge rates, respectively. Hence, for sustainable groundwater development, it is desirable to consider transient components in areas with low/high recharge and high/low discharge. Withdrawals during dry periods need to be balanced by replenishment during wet periods. However, many of these aquifers are shallow and naturally replenished but particularly vulnerable to excessive withdrawals and contamination, and protecting groundwater quality is of paramount concern.

Often, only a small part of the total groundwater in storage can be used without significant effects on surface water, ecosystems, land subsidence, or water quality. To ensure water security, currently untapped water sources may need to be developed. The assessment should take place at the national, regional, and individual aquifer scale. The effects of pumping on groundwater depletion, water quality, surface water, and ecosystems are at a tipping point for many of the critical aquifers. While in India, groundwater recharge ought to be the top priority for planning of water sector investments, it should be maximum in basins with the most intensive groundwater use and high levels of resource depletion. Instead of blindly following international practices for groundwater recharge using spreading technologies, India's strategy should be based on imaginative use of groundwater recharge of millions of existing private dug wells. Sustainable solutions at the societal level need to include people's emotional values, and views on current and intergenerational equity.

References

Datta P.S. (2014). Need for better groundwater governance for sustained water supply. IPHE National Conference on Piped Water Supply and Sewerage Systems, January 24–25, New Delhi.

Datta P.S. (2015). Traditional Rainwater Harvesting at Crossroads – Time to Seek Ethical Opportunities for Reform. Proceedings of the Aqua Foundation IX World Aqua International Congress, Reviving Traditional Water & Environment Conservation Techniques, November 26–27, New Delhi, pp. 93–115.

Further Reading

Franco, J., Feodoroff, T., Kay, S. et al. (2014). *The Global Water Grab*. Amsterdam: TNI.

Rulli, M.C., Saviori, A., and d'Odorico, P. (2013). Global land and water grabbing. *Proceedings of the National Academy of Sciences of the USA* 110 (3): 892–897.

8

Scope of Water Harvesting for Groundwater Management Strategies

CHAPTER MENU

8.1 Scope of Water Harvesting as Part of Groundwater Management

Across the world, end-users manage water resources according to different strategies and principles, depending on the amount of rainfall, potential evapotranspiration, and local cropping system (or other use of water). Although the globally aggregated volume of depleted groundwater may be very small compared to the total volume of water in the world, the impact of depleted groundwater is significant because most of the depletion is concentrated in a few areas only, where unplanned and indiscriminate pumping of groundwater has removed relatively accessible water. These areas are particularly located in the arid and semi-arid zones of the northern hemisphere and in Australia. The groundwater reserves there have lost contact with a modern active water cycle, and abstraction-induced depletion threatens long-term sustainability of the vital groundwater resources.

In areas with humid and subhumid conditions as well as semi-arid and arid conditions, broadly, four different strategies are adopted for water management.

- Harvesting excess water from high-intensity rainfall through water storage.
- Managing surface runoff from seasonal flooding through floodwater harvesting and controlled drainage.
- Improving direct water infiltration and reducing evaporation water loss; adopting soil water *in situ* conservation practices that prevent surface runoff and keep rainwater in place (e.g. conservation agriculture, level bench terraces, mulching, dew harvesting).
- Increasing water use efficiency (e.g. good agronomic practice, including use of best-suited planting material and fertility management).

The most sustainable method is to adopt a combination of strategies.

Water Harvesting for Groundwater Management: Issues, Perspectives, Scope, and Challenges, First Edition. Partha Sarathi Datta.

Hence, from the viewpoint of water harvesting for groundwater management and water security, in-depth detailed considerations are required on governance narratives; types of groundwater resources to which these narratives can apply; principles that have been included over time; and implications for sustainable development of groundwater resources now and in the future. Other aspects include combining freedom of action with sustainability principles; the dividing line between preservation and economic development of the geosphere; making relationships between geoscientists, media, politicians, and citizens more profitable; and communication and educational strategies that should be adopted to transfer the value of the geosciences to society. In this chapter, these issues are discussed to understand the possible approaches for sound governance to ensure sustained water availability in the semi-arid and arid parts of the world.

8.2 Groundwater Management and Sustainability Issues

Globally, a broad consensus definition of groundwater management for security and sustainability is difficult to achieve. For thousands of years, agriculture has been the largest user of water resources, and the most essential activity for human survival. In recent human history, water resources have been used in many other ways for different purposes. In parts of the world with rapid population growth, urbanization, and competition for economic development, demands for agricultural and other products, by-products, and their productivity have increased, creating accelerated demand and ever-increasing pressures on the limited global stock of groundwater. This has resulted in a higher rate of groundwater utilization in a very indiscriminate and unplanned manner, and water use has also dramatically changed with emerging new technologies, mechanization, specialization, and government policies that favor maximizing developmental activities.

Although these changes have had many positive effects, there have also been negative impacts, such as degradation of land, soil, and water resources at an alarming rate, loss of biological diversity, overexploitation of resources, indiscriminate waste disposal, threats of rising sea levels, worsening water scarcity, etc. However, we are now realizing that development based on the concept of exploitation and unbalanced use of water resources costs too much in terms of sustainability. The lives of the rural poor are most immediately affected because their day-to-day subsistence and livelihoods more often depend on the water resources around them. Although the exodus to urban areas has reduced pressure on rural lands, the number of people living in unsafe and overcrowded urban slums has increased, with billions of people lacking safe drinking water and basic sanitation.

By the year 2025, about 50% of the populations of developing countries are likely to live in urban areas, compared to 34% at present. An additional 3 billion people will have to be fed from a finite resource base. Unfortunately, conflicts over water resources, generated by new patterns of utilization, are generally ignored or overlooked. Using resources more efficiently and sustainably may allow us to feed the growing population with no further degradation of the environment, avoiding further encroachment on fragile ecosystems that affect natural resources. These conflicts become apparent when people challenge energy-intensive industrial technologies, because their survival depends on these technologies and doubts are arising about our ability to achieve resource sustainability with current consumption/use patterns.

With respect to renewable resources such as groundwater management, the term "sustainability" relates to establishing the longevity of natural reserves in such a way that the rates of use are co-ordinated in a balanced manner with the natural productivity rates affecting the growth and decline of the resources. Sustainability of water supply means that water resources should be used in a way that does not jeopardize water availability for future generations. Sustainable groundwater use may include consumptive use (e.g. harvesting) and non-consumptive or low consumptive use (e.g. recreation, contemplation). It can be achieved by increasing efficiency or reducing wastage or by adopting other methods. With reference to water, sustainable development strategies are interrelated with other aspects such as ethical responsibilities for maintaining and reforming institutions, changing emphasis of human understanding, behavior and practices, technological change, policies, guidelines and laws of governments, the activities of non-governmental organizations, businesses, and local communities.

In view of the growing concern about the accelerating degradation of the human environment and natural resources, and the consequences of degradation for economic and social development, sustainable management is defined as "development that meets the needs of the present without compromising the ability of future generations to meet their own needs." Sustainability relates to the continuity of economic, social, institutional, and environmental aspects of human society, as well as the non-human environment. However, the problems of sustainable groundwater management differ markedly between developed and developing countries. For instance, with respect to groundwater in developed countries, problems may arise from overuse, while in developing countries, problems are usually related to overuse, lack of management, lack of protection from depletion and degradation, and uncontrolled practices often leading to horizontal expansion areas prone to degradation.

Often, poverty forces rural people to depend on water resources for their livelihoods and survival. However, the quality and quantity of water required to sustain human life is highly debatable, due to a wide range of perceptions on what constitutes an acceptable quality of life. Indiscriminate use of water has caused a number of environmental threats, such as rising water demand and consumption, growing pressure on water systems, groundwater depletion and deteriorating water quality, pollution from point and non-point sources, and increasing deterioration of socioeconomic conditions in rural areas. If water is used indiscriminately, eventually the resource becomes scarce, sometimes to such an extent that it can no longer be used. Its price rises, resulting in increasing prices of products made from the resource. The point of non-sustainable activity is the resource use outcome which, at the extreme, means the resource is exhausted and, at a minimum, the resource system is recoverable but at a lower quality and lower rate than before. If a decision to reverse non-sustainable activity is made after reaching this point, a new starting point of "reconciliation" is created.

As the economic growth of a region is inextricably linked with the available water resources, the multipurpose role of water is complex in economically developed industrialized countries. For resource managers, it is a matter of increasing concern that the present rate of economic growth is already exceeding the Earth's capacity to produce the required resources and to absorb the pollution caused by present levels of anthropogenic activity. One strategy to counter this consists of improving land use and water resources planning with a focus on promotion of proper land use management practices; integrated waste management; reducing point source pollution by strengthening industrial and municipal waste management; laws and regulations to reduce and control pollution, and their strong enforcement; farming methods that reduce the use of agro-chemicals and fertilizers. However, numerous barriers arise because currently technological systems and governing institutions are designed and built for permanence and reliability. The real issue is the difficulty in convincing individuals, corporations, and nations to adopt the ethical practices needed, which involve change in behavior and values, at personal, corporate, collective, and societal levels.

8.2.1 The Question Remains – How to Sustain Water Supply?

8.2.1.1 Groundwater for Sustainable Water Supply

Groundwater for sustainable water supply implies the development and use of groundwater without causing unacceptable long-term environmental, economic or social consequences, through integrated, adaptive, and inclusive governance. However, none of the analyses made so far relating to aquifer-specific legal mechanisms and groundwater regulations has comprehensively assessed the norms in all relevant aspects of groundwater governance from the perspective of sustainable groundwater management. Aquifer typologies include national aquifers completely contained within state boundaries, transboundary aquifers shared by two or more states. and national aquifers hydrologically linked with international watercourses.

As described in Chapter 3, an aquifer is a permeable layer of underground rock or geological formation, which is saturated with and capable of yielding sufficient groundwater for human use. Aquifer yield can be viewed from many different scales such as for a well, for a specific aquifer or for an entire aquifer basin or system. Aquifers can be unconfined, where the water table occurs within the aquifer layer and the groundwater is in direct contact with the atmosphere through soil pores, or confined, where it is overlain and underlain by a semi-permeable or impermeable layer. A non-recharging aquifer, containing "fossil" groundwater, does not receive enough recharge on a human time scale.

The "safe yield" of a groundwater basin from the viewpoint of groundwater management is generally used to determine the long-term balance regarding the quantities of water that can be safely withdrawn annually from an aquifer system, so that withdrawals do not exceed the annual rate of recharge. However, from the aquifer dynamics point of view, this is an oversimplification, because it does not clearly incorporate other processes (such as inflows, outflows, and changes in storage) occurring in an aquifer as water is pumped from the system. Basin yield can be defined as the maximum rate of withdrawal that can be sustained by the complete hydrogeological system in a basin without causing unacceptable declines in hydraulic head anywhere in the system or to any other components of the basin hydrological cycle. An aquifer's sustainable yield must be considerably less than recharge if adequate amounts of water are to be available to sustain both the quantity and quality of streams, springs, wetlands, and groundwater-dependent ecosystems.

However, in reality, a certain amount of water always leaves the system through outflow. When large withdrawals from the aquifer continue to drain water from storage, the aquifer system as a whole shrinks because the total volume of water is less than before. Within this complex mix of competing interests, managing groundwater resources for water security requires a different time scale, and it is increasingly a challenge for scientists to address this socially sensitive issue. The response time of rivers, lakes, wetlands, and springs to groundwater pumping can be many decades, yet groundwater planning horizons are often only 5–20 years. Moreover, the volumes of invisible groundwater and the dissolved constituents in the water are highly variable. The volume of groundwater present below an area of land may vary from a few hundred cubic meters to several million cubic meters, depending on the space available. It is therefore difficult to assess in advance the volume of groundwater that already exists in a region or country, and to estimate the volume of rainwater that can be harvested to recharge groundwater in a country, a region, or within the parts of a region.

Longer pumping time frames also involve deliberate depletion of the resource, which cannot be controlled by antiquated groundwater laws and policies, which need to be updated. Therefore, groundwater should not be artificially recharged by rainwater or floodwater harvesting, exploited or managed in isolation from the overall water resources setting of the area concerned, without field studying the area extensively. It is essential to recognize these inextricable linkages and build public awareness and capacity building for water resource managers for a new vision of groundwater governance. Sustainable access to adequate supplies of good-quality water can be increased through demand management that aims to reduce loss and misuse; optimize water use by ensuring reasonable allocation between various users; facilitate major financial and infrastructural savings for cities; and reduce stress on groundwater resources by reducing unsustainable consumption levels.

8.2.2 Integrated Watershed Management (IWM)

Integrated watershed management and integrated water resources management (IWRM) are terms now widely advocated and applied. The IWM concept aims at water governance to improve both private and communal livelihood benefits from integrated technical, technological, and institutional interventions, beyond traditional rainwater harvesting, soil and water conservation, etc., to include strong institutional arrangements for collective action and market-related innovations that support and diversify livelihoods. This concept ties water governance together with water resources management in a watershed as a hydrological landscape unit with that of community and institutional factors that regulate local water demand and determine the viability and sustainability of such interventions (Liniger et al. 2011; www.wocat.net). However, the IWRM concept implies considering indirect uses and non-consumptive functions of groundwater, including the use of regular surface water generated by groundwater.

In recent times, the term "water governance" has become part of the vocabulary of professionals and academics in the water sector. Although the term has many different definitions and uses, in general it refers to the set of rules, practices, and processes in political, administrative, economic and social systems through which decision makers determine the water resources management to implement services, that influence water use, in order to determine where water flows, for what purpose, and at what cost (ecological, social, economic). Water governance is intimately linked to the physical infrastructure used for the supply/distribution of water, regulation, abstraction, storage, transport, and interaction of ecological, social, and technological systems.

For instance, in India, although under the Constitution "water" is a state subject and hence the rules/responsibilities for effective governance of water are divided between central and state authorities, there is no overall sustainable vision for water development, conservation, and management, due to fragmented approaches and lack of coordination among departments, implementers, and regulators. The National Water Policy (NWP), formulated in 1987, revised in 2002, and debated in 2012 as a draft NWP, seems to be indifferent on a commitment about "right to water" as distinct from "water rights." The validity of the per capita water availability index may not be so relevant in relation to the socioeconomic disparities in water usage.

The water sector public services reform programs focus on policy, legal, and administrative realms but overlook governance arrangements, which are often a central issue in Indian politics. Policy discourse has only recently started to consider governance of water supply as a key issue. The large infrastructure market and growing need for water supply intervention attract the interest of international development agencies, multinational corporations, and investors. Private sector involvement as a proposed mode to address this problem is politically contentious, but not seriously contested at a policy level. However, both private sector and donor investment in the water sector have been falling over recent years. Consequently, better knowledge and governance are required in reform strategies and alternative modes of supply. From a governance policy perspective, alternatives to official water supply are a feature of daily life for many in India, but there are huge knowledge gaps.

The widespread use of groundwater is a vast supplement to the water consumed, and is important for financial, economic, and ecological sustainability of water supply; however, precise figures are hard to obtain. Uncertainty about groundwater resources is usually commonplace when dealing with water policy-related problems, particularly the general exaggeration on predicting future water demands. An important step in trying to manage groundwater resource is to assess the seriousness and type of the assumed problem. Groundwater use is mediated or subsidized by economic and political power via social and technological systems; booster pumps, filtering facilities, and tankers etc. to keep tariffs very low by not increasing coverage. However, this is more readily available to more powerful groups, which leads to depletion in the resource and negative effects for poor users and the hydrological system as a whole.

Often, the adverse effects of "overexploitation" are misunderstood or exaggerated. This is often the case in relation to the interpretation of a long-term (e.g. 10 years) water level decline as an indication of groundwater abstraction higher than the average renewable resources. As previously explained, such a decline may be due to (i) dry spell, (ii) transient situation, or (iii) scarce or incorrect data about groundwater levels, stream flow, climatic conditions, groundwater abstractions, and natural recharge. These two factors are usually difficult to ascertain in arid and semi-arid parts of India because it is necessary to have adequate funds in order to obtain better data in regard to quantity or quality of information. Moreover, the natural recharge in semi-arid regions will only be accurately known after a substantial number of years of good climatic and hydrological data have been collected.

8.2.3 Approaches to Water Governance

Globally, for water management, water governance generally adopts a variety of approaches. Two broad approaches are (i) more instruments oriented, targeting governance arrangements and processes, which seek to enhance efficiency, equity, and effectiveness of water management, and (ii) critical analysis of governance processes, and the extent to which prevailing processes result in equitable access to resources and services. Therefore, understanding a particular water governance configuration requires recognition of the interdependency of prevailing social, technological, and ecological systems. Hence, context-specific historical, social, political, and cultural processes are important within water governance at national and regional level.

8.2.4 Historical Approach to Water Management

8.2.4.1 Command and Control

Since the late 1970s there has been growing awareness of problems associated with various sociopolitical aspects of water management. The dominant management approach has been "command and control," which is centered on technological solutions and regulations. While technologies such as water meters, supervisory control and data acquisition (SCADA), GIS, and SIM cards are considered important in the design of water supply systems, in most developing countries there is lack of monitoring of the performance of such technologies. Moreover, this approach has been found to fail on several fronts, due to the following reasons.

- *Fragmentation* across geographic, sectorial, administrative, institutional, and disciplinary boundaries.
- *Exclusiveness*, with a technocratic management of top-down style, side-lining some stakeholders.
- *Inflexibility* due to massive infrastructural investments being made using projective/predictive planning.
- *Reactiveness* as a result of overcompensation for inflexibility.

In the face of complex problems, such as anticipated climate change and rapid urban development, new water management approaches are needed in order to overcome the shortcomings of the "command and control" approach. However, over the past few decades, it has been observed that approaches such as integrated to overcome fragmentation; participatory to remove exclusivity; and adaptive to check inflexibility could not provide fool-proof solutions. Recognizing the importance of sociopolitical and governance considerations, anticipatory water governance needs to be developed to address reactive behavior.

8.2.4.2 Anticipatory Water Governance

A broad definition of the anticipatory approach is: "Develop a system of institutions, rules, strategies, and norms for resilient water services that provide a way to use foresight for the purpose of reducing risk and to increase capacity to respond to unforeseen events at early stages instead of at later stages, without generating inflexibility." However, paradoxically, it has been observed that anticipatory water governance usually tends to be associated with predictive long-term planning and infrastructural investments, which in turn limit the capacity to adapt. On the other hand, water governance that relies heavily on adaptive capacity tends to be characterized by reactive behavior in practice, lacking vision and with limited capacity to anticipate. An adaptation can be made before, during, and/or after by anticipating and/or responding to a stimulus. Hence, there is a need to operationalize the concept for practical application, to improve resilience with a holistic approach, by balanced consideration of integrated, participatory, adaptive, and anticipatory approaches.

8.2.5 Principles for Governance of Groundwater

All the commonly accepted water governance principles are applicable, which include integrating sustainability, equitable access, accountability, transparency, participation and representation, with overall water resources management, However,these need to be adapted to the specific character of groundwater and supplemented by three principles of special relevance to groundwater: precautionary environmental principle (i.e. protecting aquifer water quality and assuring recharge), economic principle, and knowledge management principle.

8.2.5.1 Precautionary Environmental Principle

This principle aimed at basin-level management facilitates sustainable development through (i) resource-based management, (ii) preventing risk/hazard (including overexploitation and pollution), and (iii) notification procedures. It requires the use of hydrological rather than administrative units for groundwater management, and also facilitates management at the lowest appropriate level and conjunctive use and management when groundwater is appropriately included in the scope. However, this principle is limited and hardly used, because at the global level, groundwater resources outside transboundary aquifers/basins are not fully included. Since politics is a function of administrative units and not aquifers and river basins, this principle presents a practical challenge. Many environmental principles depend on and/or support each other. This principle would prevent the adjoining countries/states from significantly impacting other aquifer countries/states and would require them to implement pollution prevention principles to maintain groundwater quality and prevent overabstraction throughout the transboundary aquifer.

However, in practice, for the management requirements there is a lack of aquifer-specific data regarding the quality and quantity of groundwater. Therefore, the precautionary principle is highly relevant, and the legal approach to management should include measures that protect the resource from overexploitation or irreparable contamination. Since there are no environmental principles which explicitly deal with overabstraction, many states may lack the practical and legal means to prevent significant harm to groundwater resources. Actions such as diverting a watercourse that recharges an aquifer or abstracting groundwater may affect the groundwater resource. The principle of notification of planned measures would require that the potentially affected states receive advance notification. Countries/states are also obliged to notify other aquifer states during emergencies such as droughts.

8.2.5.2 Economic Polluter Pays Principle

This principle require polluters to internalize the costs of pollution, although still appearing sporadically in governance at global level. Interestingly, it is not always paired with principles for preventing pollution/harm or protecting ecosystems, principles which it supports. Since many products traded internationally use groundwater in their production processes, the principle of promoting an open international economic system is important in groundwater governance. It facilitates economic growth, ensuring that environmental trade policies are non-discriminatory and do not constitute a "disguised restriction."

Although water as an economic good and maintaining an open international system are not addressed in global water governance, these principles are implemented *de facto* in trade regimes, leading to increasing demand for water-intensive products, which passes on economic and environmental costs to water users. In other words, the costs of these activities are simultaneously incorporated into water prices but excluded from prices of groundwater-intensive products across the world, possibly shifting financial burdens and affecting implementation of rights-based approaches and ecosystems protection.

8.2.5.3 Knowledge Management Principle

Governance and management need to be realistic, based on understanding of both hydrogeology and socio-economy as well as political realities, pulling together information to establish stakeholders' interests, identifying a lead agency, defining rights and incentives, and ensuring transparency and accountability. Hence, a normative framework for sustainable groundwater management requires that groundwater governance (i) use an appropriate common terminology that is rooted either in the state-of-the art hydrogeology or legal norms; (ii) include definitions and scope that recognize the groundwater being both part of and apart from the contemporary hydrological cycle, thus including aquifers of all types whether non-recharging, layered, or linked to surface water; (iii) include norms presently underrepresented in legally binding texts; (iv) reconcile tensions between principles; (v) elaborate best practices for well-accepted principles that are most challenging to implement; and (vi) introduce principles or mechanisms to cope with the effects of trade and climate change.

8.3 The Missing Elements in Groundwater Management

Water governance issues are the responsibility of a multitude of government bodies, and therefore recommendations for reforms from different initiatives fail to be implemented due to lack of broad consensus and interministerial agency co-ordination. The different types of risks pose a grave danger to agriculture and the livelihoods of millions in the Ganga basin. The direct benefits of recovered resources from waste water could make an economically attractive case for the private sector to focus on waste-water treatment and management. However, whichever approach is adopted, the management of water resources must be based on the above-mentioned inextricable linkages that are especially aggressive to the depletion and degradation and on adequate knowledge of the situation of the groundwater system and its replenishment.

8.3.1 Critical Knowledge Gaps

Some of the important matters of concern to planners and managers are gaps in knowledge of groundwater rechargeability, its quantity and quality, the location of recharge intake areas, the interaction between groundwater and surface water, sources of pollution, pollutant dynamics, etc. The measurements of water table fluctuations and pumping tests do not provide all the information needed to bridge these gaps, and most groundwater development research has been fragmented and technocratic, related to groundwater flow and remediation. Moreover, there is no way to evaluate the maximum exploitable groundwater volume in advance or in absolute terms, and highly technical knowledge of the aquifer systems is of limited use for practical management purposes. Unfortunately, public demand and use of groundwater in the socioeconomic context, and the consequences of these items, have been inadequately characterized.

More research is needed on the dynamics of pollutants in the groundwater and its attenuation capacity for pollutants under natural and exploited conditions, based on a well-designed monitoring network. Pollution sources should be identified and strategies developed to contain the spread of pollution from known sources and to develop groundwater vulnerability maps. Based on these maps, potential groundwater recharge zones and protection zones need to be clearly delineated with land use changes to restrict/eliminate unplanned waste disposal and agro-chemical application in these areas. So far, no practical solution exists to account for water and energy efficiency at the withdrawal point. Further research on economic aspects of global groundwater governance and the relationship between the global framework and domestic rules and rights would greatly support such an endeavor.

India's water resources have both internal and external components (i.e. entering India via transboundary waters), resulting in high dependency and uncertainty about future water availability. A large part of the country also suffers from saline water intrusion into aquifers. Industries consume significant amounts of groundwater and create large volumes of polluted discharge. While discussing desalination to address water scarcity, only areas near sea water are likely to benefit. But it is extremely capital- and energy-intensive and a few years later a surge in water recycling and reuse is seen: once the water is desalinated, it is cheaper to keep it clean than it is to make more.

8.3.2 Tackling Groundwater Pollution

From the viewpoint of assessing water supply for meeting various social, economic, and environmental objectives, integration of data on groundwater quality with data on water supplies is very important but is rarely undertaken. In order to prevent groundwater quality deterioration, the first step is generating reliable and accurate information through water quality monitoring to understand the actual source/cause, type, and level of contamination. However, existing methodologies are inadequate to identify the various sources of pollution and there are very few observation stations in the country that cover all the essential parameters for water quality and hence the data obtained are not decisive on water quality status. Moreover, groundwater quality monitoring involves expensive and sophisticated instruments, which are difficult to operate and maintain, and require substantial expertise in collecting, analyzing, and managing data. Also, in the absence of any stringent norms on water quality testing, results can change across agencies depending on sampling procedure, time of testing, and instruments used.

To prevent groundwater quality degradation from sea-water intrusion, artificial recharge techniques are available for different geohydrological settings. Artificial recharge of groundwater can push the sea water–fresh water interface seawards. These techniques can also be used to reduce levels of fluoride, arsenic, or salinity in aquifer waters on the principle of dilution. However, the issue is availability of good water for recharging the large extent of contaminated aquifers in arid and semi-arid regions. For industrial pollution, the issues are of three types: pumping out polluted water from the aquifer; treating this water to safe limits; and replenishing the depleted aquifer with fresh water. Due to the presence of highly toxic substances in trade effluents, technically feasible methods to clean polluted water often do not exist, and may be economically unviable in the Indian context.

In India, the Central Pollution Control Board and the state pollution control boards are the pollution watchdogs. However, monitoring of groundwater quality has come under their purview only recently, and does not cover non-point pollution from agriculture, and the network of monitoring stations is not dense enough. There is a paucity of civil society institutions capable of carrying out such challenging, technologically sophisticated, and often politically sensitive tasks. This would be enhanced by better knowledge and information about the nature of groundwater pollution, potential sources of threats to water quality and degrees of vulnerability, the ill effects of using contaminated water, and possible preventive measures.

The 2017 budget provided an impetus to irrigation in India, with a 28% increase in budget allocation for the Pradhan Mantri Krishi Sinchayee Yojana, a national mission to improve farm productity, through an additional provision to the long-term irrigation fund set up by the National Bank for Agriculture and Rural Development (NABARD). Outlays for river basin management (Namami Gange) have been increased by 5%. Further, Smart City missions are likely to spend thousands of crores of rupees. Although the government has introduced policies mandating reuse of treated water from sewage treatment plants in nearby thermal power plants, strict implementation is the key. With no major revenues being accrued by government agencies from domestic water supply services, any additional investment for provision of safe water for drinking and other purposes is likely to induce unprecedented financial burden.

8.3.3 Gaps in Legal Frameworks and Their Application

Although there exists some legislation on groundwater, it is scattered across several instruments and often appears "theoretical," partial, inconsistent or outdated, poorly adapted to the realities, and hard to implement. Legislation on groundwater quantity is usually separate from legislation on quality, which can be an obstacle to management. Customary rights are somewhat ignored in legislation. This is likely to lead to problems in application, and to negative impacts on people. Many countries have reserved legal ownership of groundwater to the state, to counter private overexploitation. However, local people generally assume that they, and not the state, own the groundwater, which leads to unsustainable management.

8.3.4 Gaps in Goals, Policies, and Plans

Due to limited awareness of issues and the short time horizon of politicians and decision makers, policies on groundwater are somewhat incoherent or even non-existent, without the long-term vision needed to link groundwater governance to well-defined societal goals of growth with equity, sustainability, and efficiency. Policies to reduce groundwater pumping by increasing irrigation efficiency sometimes had the opposite effect, as farmers expanded the irrigated area, and increased conveyance efficiency contributed to lower groundwater recharge. Where subsidies have been used to restrict groundwater development in already overexploited areas, this has tended to freeze existing patterns of rights and to restrict access by the poor.

8.3.5 Gaps in Performance of Public Agencies and Stakeholders

Capacity and performance of government institutions/organizations are variable and they perform poorly with fuzzy mandates, scant staff and human capacity, limited political support or institutional authority, and inadequate budgets. Fragmentation and lack of clarity on responsibilities among agencies are a common problem. Agencies tend to take a top-down engineering approach, whereas the challenges of the complex socio-economy of groundwater require also a complementary bottom-up stakeholder involvement approach. The stakeholders generally play a limited role in a remarkably passive manner, partly due to lack of awareness or knowledge, but mainly because the institutional structures for participation are not in place.

In India, industry, production, the economy, and air and water pollution have grown in parallel. Water co-exists with all kinds of substances, added variably by industries, ecosystems, and communities that use and surround it. It is subject to constantly changing weather conditions, as well as the infrastructures and geological features through which it flows. Many by-products – salts, oils, metals, plastics, drugs and chemicals, dyes, etc. – make their way into water. These issues tend to manifest in a variety of ways. In areas where drinking water supplies share the same space as waste deposits, a range of health problems such as skin lesions, dysentery, reproductive disorders, neurological conditions, cancer and death begin to emerge. Ecosystems change, fishing villages dwindle, and livelihoods are altered.

Although an increase in financial resources for the irrigation sector is a positive step, it is important to prioritize water provisioning and not just infrastructure creation. Access to irrigation is limited, even in water-rich states with built irrigation infrastructure, due to poor electricity availability or poor maintenance of irrigation pumps or damaged secondary and tertiary canals etc. India's future energy goals such as the ambitious target of 175 GW of renewable energy by 2022 and developing ultra-supercritical coal technologies could also be constrained by availability of water. The government's Swachh Bharat scheme is also co-dependent on access to water. The National Sample Survey 2012 revealed that among rural households, only 40% are connected to sanitation facilities and 46% have access to drinking water within their premises. River action plans and policies need to shift focus from sewage treatment to a more holistic management strategy and also include focus on other river basins.

Under the above-mentioned circumstances, it would be unethical and unfair to turn a blind eye to indiscriminate water extraction and focus exclusively on rainwater harvesting, water conservation, etc. for groundwater management, without a very clear understanding of the pros and cons of these approaches, based on secondary information on benefits. To effectively implement any program, groundwater management at the local level does not exist. Hardly one-fourth of the total area of the aquifers has been mapped so far, which is insufficient to make a visible impact on aquifer management (including their recharging) along with stakeholder participation. With multiple ownership of wells and overlapping jurisdictions of common groundwater resources, an effective legislative and policy response to address this problem has yet to take place.

References

Liniger, H.P., Mekdaschi Studer, R., Hauert, C., and Gurtner, M. (2011). *Sustainable Land Management in Practice – Guidelines and Best Practices for Sub-Saharan Africa*. Rome: TerrAfrica, World Overview of Conservation Approaches and Technologies (WOCAT) and Food and Agriculture Organization of the United Nations (FAO).

Further Reading

Datta P.S. (2013). Groundwater vulnerability to changes in land use and society in India. In: Understanding Freshwater Quality Problems in a Changing World, Proceedings of H04, IAHS-IAPSO-IASPEI Assembly, Gothenburg, July (IAHS Publ. 361, 2013), pp. 345–352.

Datta P.S. (2014). Need for better groundwater governance for sustained water supply. IPHE National Conference on Piped Water Supply and Sewerage Systems, January 24–25, New Delhi.

Datta P.S. (2015). Traditional rainwater harvesting at crossroads – time to seek ethical opportunities for reform. Proceedings of the Aqua Foundation IX World Aqua International Congress, Reviving Traditional Water and Environment Conservation Techniques, November 26–27, New Delhi, pp. 93–115.

Datta, P.S. (2017). *Better Groundwater Governance Only Can Ensure Sustained Water Supply*. Germany: Lambert Academic Publishing.

Datta Partha Sarathi (2013) Geoethics and geoscientists role to protect geodiversity for societal development. GEOITALIA2013, Session Geopolicy – L 2 Geoethics and Society: Geosciences Serving the Public, September 16–18, Pisa.

9

Guidelines to Make Water Harvesting Helpful and Meaningful for Groundwater Management

CHAPTER MENU

9.1 Groundwater Augmentation Approaches

In order to augment groundwater resources, two approaches are usually adopted: direct (by artificial recharge) or indirect (by induced recharge). Artificial recharge is a technique to control surface water (the source to be used) and to augment the exploitable groundwater resources artificially. Depending on the particular case, it allows storage of water in arid areas where control of very irregular surface runoff (erratic floods) is difficult and where protecting stored water against evaporation is most needed. Induced recharge is fairly widespread in exploited zones of alluvial aquifers along perennial streams, and near land depressions collecting surface runoff. Depending on declines in the groundwater level, the water exchange may be surface water to groundwater or groundwater to surface water.

In recent years, the term "artificial recharge" has been interchangeably used with "managed aquifer recharge" (MAR), by also including induced recharge and sometimes also the recovery of stored water. Artificial recharge can help restore the imbalance of an excessively exploited aquifer, or create surplus, which could allow more groundwater to be withdrawn than is warranted by the natural recharge, and in some cases primarily intended to control/improve water quality. It can also be useful in the waste-water reuse chain, by facilitating storage of waste water and its purification.

Artificial recharge can be subdivided into extensive and intensive techniques. Extensive techniques (ponding, infiltration basins) are cheapest and better adapted to arid conditions, but require more space. Intensive techniques (shallow well, shaft, pit or deep well injection) are more expensive, less durable (due to clogging), and more

Water Harvesting for Groundwater Management: Issues, Perspectives, Scope, and Challenges, First Edition. Partha Sarathi Datta.

localized. However, whichever approach is taken, clogging of the streambed, riverbanks and depression beds and pollution of surface water and groundwater may become limiting factors in the longer term.

9.1.1 Issues and Concerns for Meaningful Rainwater Harvesting

The strengths, weaknesses, opportunities and threats (SWOT) analyses of rainwater harvesting systems (RWHS), as described in Chapter 6, suggest that in most water-scarce regions of the world, including India, rainwater harvesting offers extremely limited potential to reduce the imbalances in water demand and supply, because a significant part of such regions is characterized by low mean annual rainfalls and high inter-annual variability in rainfall with high potential evaporation (PE). A larger part of such variability occurs during the rainy season, increasing the occurrence of hydrological stresses and reducing the runoff potential. In many regions with medium rainfall and medium to high PE, poor groundwater potential poses constraints for recharge. The overall potential of RWHS would be high in regions with high rainfall and medium evaporation. For instance, in India, a large part of water-scarce regions which fall under the medium rainfall and medium to high evaporation regime are underlain by hard-rock formations such as basalt, crystalline rocks, and other consolidated formations such as sandstones. In such areas and also in areas having silty-clay and clayey soils, RWHS are likely to have low efficiency, and planning of recharge schemes should consider surface water impoundment of all the available excess flows, rather than direct recharge.

Worldwide, as rain and precipitation trends and patterns have changed over the years, both surface and groundwater resources are less predictably replenished. In situations where droughts, floods, and storm conditions are on the rise, water use becomes regulated and rationing common. Everything from agriculture and industry depends on reliable infrastructure. Aging infrastructure undermines dependable storage, treatment, and delivery of water so the choice for only water harvesting and rainwater harvesting for groundwater management depend both on time and on the state of the resource. Small-scale "traditional" RWHS cannot be widely implemented because WH may not always be appreciated by local communities, if it does not lead to better access to water or increased harvest. Moreover, depending on the characteristics and distribution of water resources in any region, as the interest in water rights increases, tensions involving powerful individuals and groups rise.

9.1.1.1 Water Resource Characteristics and Distribution

Water resource characteristics and distribution differ from region to region and within the parts of a region. In many water-scarce regions, water demands far exceed the supplies, and due to vulnerability to hydrological stresses, exogenous water is required. Enhancement of safe water availability and supply will be determined by the policies, plans, and technologies available, in addition to political, socio-economic, and other factors.

9.1.1.2 Effective Implementation of RWH Strategies

Effective implementation of RWH strategies for management of water supply and demand at local and regional level require that each area/region be investigated separately, with a more systematic monitoring network and research on pattern of water use from different sources, availability of different types of water resources, natural recharge to groundwater from rainfall, dynamics of pollutants in groundwater flow under natural and exploited conditions, etc.

9.1.1.3 Evaluation of Water Harvesting and Groundwater Recharge Systems

Evaluation of water harvesting and groundwater recharge systems poses several direct economic and ecological or environmental complexities due to the difficulty in quantifying the hydrological impacts and the various economic costs and benefits, such as inflows, storage and recharge efficiency, and economic value of the incremental benefits. In many water-scarce basins, a strong tradeoff exists between maximizing the hydrological benefits from RWH and making them cost-effective. In many water-scarce basins, RWH interventions lead to distribution of hydrological benefits, rather than augmentation. There is an optimum level of water harvesting which a basin can undergo to help optimize the gross value product of water in relation to economic, social, and environmental outputs basin-wide. Therefore, the economics of water harvesting cannot be determined for structures based on their individual benefits and costs when the basin has limited surplus water; rather, it should be done on the basis of incremental benefits. The higher the degree of basin development, the higher would be the marginal cost and the lower the marginal benefit from water harvesting.

In basins which experience high interannual variability in stream flows, the tradeoff between hydrological impacts of water harvesting and economic benefits is likely to be large. With increasing storage capacity of RWH systems, economic viability becomes poorer as the average cost of water harvesting per unit volume of water increases. In closed basins, there is a tradeoff between local benefits and downstream benefits. Upstream diversions reduce the prospects of storage and diversion systems downstream. In many basins, there is an apparent tradeoff between maximizing overall benefits for basin communities in terms of enhancing the gross value product of water, and maximizing the local benefits of water harvesting in upper catchments. This is because in these basins, water from well-endowed regions with low water demands is being diverted to poorly endowed regions with high water demands, enhancing its social and economic value.

9.1.1.4 Scale Considerations

Scale considerations are also extremely important in evaluating the cost and economics of water harvesting, almost everywhere where high interannual and interseasonal variability in rainfall exists. Sometimes, potential social benefits of improved regional equity may influence political decisions to intensify upstream water harvesting, even if this may reduce net benefits at the basin level. In developing countries such as India, RWH approaches made so far have made little or no effort to involve spatial and temporal scale considerations when analyzing the physical and economic impacts of water harvesting. Issues for concern include (i) optimal level of water harvesting in different river basins that averts unintended downstream impacts and (ii) ecological and environmental impacts of water harvesting in terms of reduction in environmental flows, or increase in lean season flows in different hydroecologies.

A major challenge of global water management is to find approaches to address socioecological controversies in a socially just and ecologically effective manner. Water governance dilemmas are multifaceted and constantly changing, and therefore a flexible analytical apparatus is needed to appreciate emerging themes and enablers for positive change. While the case studies in Chapter 5 were diverse in terms of their geography (i.e. Africa, Australia, Europe, North America, and Asia), scales (i.e. local dilemmas to transnational issues), and specific water dilemmas (e.g. flooding, pollution, and agriculture), they help to illustrate some of the underlying perspectives and implementations that underpin promising configurations in water governance, aiding our understanding of which innovative arrangements might help to promote more adaptive forms of transformation.

9.1.1.5 Water Issues and Dilemmas

Water issues and dilemmas are inherently products of complex political and socioecological entanglements, and therefore solutions will not emerge with from technological or economic approaches alone but rather by paying attention to political dynamics. It is desirable to take into consideration the various actors' actions, to help facilitate collaboration and co-ordination that allow for transformations to take place. For surface and groundwater, it is critical to characterize the quality of the water and variations in chemistry throughout the year caused by seasonal climate changes, seasonal agricultural and industrial activities impacting runoff and demand, and it is also necessary to determine flow patterns, which vary throughout the day, week, and seasonal. It is strongly recommended to perform a proper pilot study to evaluate the proposed water source for the application(s) intended to minimize the risk of failure and identify potential problems.

For instance, in India, of the water available, over 80% is used for agriculture, but irrigation efficiency is limited. The Ganga basin covers more than 25% of India's geographical area and is inhabited by over 450 million people, with 60% dependent on agriculture for their livelihoods. Groundwater, on which more than 60% of agriculture and 85% of drinking water is dependent, should be given attention on a par with surface water. Since rainwater and groundwater are two interdependent water sources, farmers who use the alternative water source create an additional negative impact on groundwater users, by diverting part of the recharge of the aquifer. Farmers may use various long-term irrigation strategies: irrigation relying only on groundwater, on rainwater, and on a combination of both water sources, and strategies depend both on costs and qualitative considerations.

9.1.1.6 RWH Implementers to Deal with Water Supply Challenges

Water use is ubiquitous, and the range of technology and expertise is also vast all over the world. However, the problem is finding the right technology or expertise, and precisely when and where that expertise is actually needed. One difficulty is how to bring all the actors together quickly and effectively. Due to the fact that efforts are fragmented, it is not easy to find out what the options are, where breakthroughs have happened, who is dealing with which challenges, and who might know how to fix them. Concerns are being raised, calling the politician–bureaucrat–contractor nexus the biggest bane in water sectors around the world, including India. Most people generally have very little idea about how water comes to them and who owns it. Most only know that water is supplied by the utility services.

Rainwater harvesting for groundwater governance and management should also focus on access to safe quality and sufficient water for reviving the agricultural sector, improving public health and strengthening the rural economy, particularly in India's context. Therefore, more efficient irrigation methods must be adopted, especially in the water-scarce regions of the country. To protect and improve the groundwater level for sustainable yields in bore wells, in India, for example, the government made it mandatory to construct rainwater harvesting or conservation structures on all premises exceeding a plot area of 200 m^2. However, owners of residential buildings, which are expected to bear the cost, have been reluctant to set up percolation pits for lack of space. Moreover, costs involved in the accurate construction and regular maintenance of RWH facilities have also discouraged residents. RWH structures should be properly fenced to avoid accumulation of waste, and the top layers of pits should be periodically cleaned and replaced to prevent silt accumulation. In addition, for most facilities, observation wells are also essential to monitor water levels and quality.

Therefore, the policy signals need to be analyzed through the lens of water especially for key areas: irrigation, access to drinking water and sanitation, and river basin management. To ensure that rainwater harvesting can work for groundwater management for the benefit of all, the greatest challenge is better governance to shift the focus from funding technological solutions to changing human behavior in water demand, use, and consumption patterns, and raising public awareness about RWH and groundwater occurrence, availability, rechargeability, quantity and quality, and then applying RWH in an equitable manner, valuing water as an economic resource.

9.1.2 Possible Options for Rainwater Harvesting for Groundwater Management

- Ascertain whether and where groundwater needs to be used for drinking supplies or high conservation ecological values, in relation to environmental values of the aquifer and intended uses of protected water.
- Transport water effectively from areas of plenty to areas of scarcity. In practice, this presents a range of financial, political, and engineering challenges.
- Use knowledge and information in awareness-raising efforts to encourage political and stakeholder participation and support in the local context on key issues that can help to catalyze data acquisition, processing, and interpretation.
- Create more systematic governance frameworks, based on success stories, to link rainwater harvesting to groundwater depletion and degradation, which will improve social awareness and acceptance of the need to act.
- Regulate on the provision of water rights and provide strong enforcement of regulations covering land encroachment and water grabbing, which can be powerful instruments of groundwater governance. However, there may be limited acceptance of this approach by users, and it may be difficult to impose regulations and sanctions.
- Establish policy and planning linkages with interrelated sectors, and mandate the preparation of area-specific management plans. Organizational reforms can greatly improve water governance, by clarifying and consolidating mandates, empowering agencies and bringing management to the lowest feasible level.
- Increase private sector involvement to strengthen governance, particularly in the sharing of knowledge and expertise in abstraction and pollution control.
- Create an integrated system of adequate groundwater supply, based on spatiotemporal variation in different time scales of groundwater recharge, because groundwater

withdrawals generally have both recent and past recharge components; the present to past recharge ratio is large in humid areas and small in arid areas, and transient and steady-state conditions prevail at low recharge rates and at high recharge rates, respectively. Hence, for sustainable water management, it is desirable to consider the transient component in areas with low/high recharge and high/low discharge (Datta 2013b).
- Encourage change in public attitudes and behaviors by developing coherent long-term policies to control communication by the mass media of incorrect information and to discourage regional polarization from social and political angles with parochial interests (Datta 2005, 2013a, 2015b, 2016, 2017).

The most important best practice for making RWH meaningful for groundwater management would be better governance to bring change in behavior to improve inter-agency co-ordination for reversing water scarcity trends, and the successful attainment of reforms in working towards sustainable water supply, for both the present and the near future. Behavioral changes (in this context) involve better governance to harness users' inherent potential to enhance efficiency by individual knowledge, transparency, trust, respect of interests, share of resources and responsibilities, and to educate the public about the water quantity that can be conserved by following social norms, through water-focused awareness campaigns, workshops, education, and professional training (Datta 2016, 2017). However, in a society not amenable to discipline, awareness alone may not be enough to bring changes in water users' attitudes and behaviors.

9.2 Behavioral Change for Groundwater Governance

Water supply management is not just about new infrastructure and technologies, but increasingly about actual water resource assessment, water demand management, effective water allocation policies and incentives. It is about building systems that are broadly resilient to drought and scarcity, managing demand as well as improving supply. Politicians, decision makers, and managers need to increasingly co-ordinate dedicated strategies and action plans at national and subnational levels. To make progress, water needs to be connected to the major interests that drive economies: agriculture, energy, and the urban, industrial, and service sectors. Reviewing existing public water policies, regulation and investment strategies for the Sustainable Development Goals (SDGs) and adaptation to anticipated climate change will be critical.

Groundwater problems in many countries, including India, are also due to the omnipresence of indiscipline, aggressiveness, corruption, "isms" of caste and creed, and politicization of governance organizations, which have significantly eroded the integrity, efficiency, and efficacy of the civil service structure. While people elect governments, very few governance systems work for the people. Barring exceptions, those who are elected to democratic institutions are neither trained, formally or informally, in law making nor do they seem to have an inclination to develop the necessary knowledge and competence in their profession. At times, it seems that the certainty of corruption in governing structures has heightened opportunism and unscrupulousness among political parties, causing them to seek opportunistic alliances and coalitions, often without popular mandate. Yet, they remain in power due to inherent systematic flaws.

9.2.1 Lack of Accountability

In many parts of the world, the manifestos, policies, and programs of political parties have lost meaning due to lack of accountability. Social and non-governmental groups, if they rise above their partisan bias, can play a vital role in expecting accountability from every organ of governance. It is imperative to improve governance in order to monitor the performance of the executive, judiciary, and legislature to ascertain the level of popular satisfaction or dissatisfaction with the governance/authority concerned.

9.2.2 Regulatory Measures

Measures to regulate exploitation of groundwater are mainly achieved through control of borehole drilling. To manage the situation, *ad hoc* fragmented and tactical approaches have been generally taken, because only about 25% of the aquifers in India have been mapped so far. Since no method exists to evaluate in advance or in absolute terms the maximum exploitable groundwater volume at local/regional scales, and the hydrogeological, social, economic, cultural, and political factors vary greatly, no single management template can be developed from the enormous number (~30 million) of individual pumping decisions. This mandates linking water management with changing use, to harmonize medium- and long-term actions, balancing groundwater extraction with recharge estimated on the basis of the intensity and distribution of seasonal rainfall and also by improved direct methods described in Chapters 2 and 3, and revising and reconsidering all such estimates from time to time (Datta 2005, 2013a, 2015b, 2017).

In many cases, waste water from treatment plant processes is discharged into rivers, other bodies of water, or back to the treatment facility. Such discharge should be regulated and a pollution discharge elimination system permit should be made mandatory. In some cases, contaminants in waste water can exceed the regulatory limit due to concentration in processes such as cooling towers. Therefore, proper monitoring of the chemistry of waste water and recycled water is necessary to ensure permit violations do not occur.

A precise understanding of the groundwater system in different regions is desirable, focusing more efforts on assessment of the variability of water resources according to exploitable resources at the watershed level. Integrated multidisciplinary science-based research efforts must be promoted to establish a sound and sustainable structural relationship between the water cycle and human activities. For climate-resilient management, creating an integrated water supply system based on better assessed spatial and temporal variation in recharge time scales may be useful, by integration of high-quality data and data assimilation schemes.

9.2.3 Role of Companies and Corporations

While it has been realized by many forward-thinking companies that the future of fresh water can be improved through co-ordinated action, and they are creating healthier water sources and diverse economic opportunities and have developed strategies to reduce operational water use, until recently few had developed mechanisms to address larger water challenges such as restoring flow to depleted groundwater. Companies need to take responsibility for the water needs of communities and supply chains, to help build water security by shrinking the human water footprint and restoring health to fresh-water ecosystems. Many companies are trying different ways to raise awareness, influence policy, and initiate projects that build long-term water security for communities, businesses, and the environment.

The most realistic measures would include:

- reduce use and protect groundwater by a drastic decrease in consumption
- reduction of waste-water discharge into the hydrological system
- use of salt and brackish waters
- use of water storage in underground aquifers of river floodplains.

For climate-resilient groundwater governance, demands on groundwater and the consequences of these demands have to be characterized. Groundwater exploitation has to be achieved through regulated control,and reliable volumetric assessment of available groundwater, its scope for augmentation, distribution, reuse/recycling, its existing depletion and degradation, and its protection. A holistic view needs to be taken, considering issues such as urban growth and changing land use, identifying pertinent parameters, phenomena, processes, and possible changes in the hydrological cycle while simultaneously addressing policy, based on reliable socioeconomic analysis.

At the launch of the SDGs in 2015, almost every government, large non-government organization (NGO), corporation, and United Nations body signed up and celebrated them. Having pursued the strategy of GDP growth with single-minded recklessness for many decades, the planet's biocapacity each year is now vastly outstripping the ability of nature to absorb anthropogenic waste and replenish the resources being used. Every new development degrades soils, pollutes water, poisons air, and exploits human beings.

In order to achieve better governance and to avoid the negative aspects of groundwater exploitation, the following suggestions may possibly help.

- Governance can play significant role in the planning and integrated management of water resources, through perception change in understanding surface water and groundwater, and their relationship with land use planning. To co-ordinate these goals, water administrations need to improve their hydrogeological capabilities much beyond what has been seen in recent decades.
- In addition to large-scale planning, governance has to undertake immediate groundwater decision making at the local scale through proper institutions for groundwater management, establishing effective participation of all users or stakeholders. For better governance, all users or stakeholders should be given adequate hydrogeological training and knowledge, so that they achieve a better understanding of groundwater dynamics.

9.2.4 Behavioral Change for Perception Management

Since the perception of the groundwater situation among stakeholders varies significantly due to lack of understanding and knowledge, the consequences of groundwater abstraction for a specific location may be quite different. The owners of wells usually pay for construction, maintenance, and operation, but do not usually pay the external costs caused by the impacts of groundwater abstraction. There exists a general consensus that, in order to avoid conflicts and to move from confrontation to co-operation, water development projects require the participation of the social groups and stakeholders affected by the projects. Participation should begin in the early stages of the project and should be, as much as possible, bottom up and not top down. Therefore, the stakeholders should be clearly identified first, and then the time and place for their inclusion in decision making should be decided.

In some arid and semi-arid areas of India, in dealing with the ecological impacts of overexploitation, the influence of conservationist groups may be weak compared to the influence of farmers' associations or urban water supply companies. The necessary participation of the stakeholders demands that they are made aware of how the issue may affect them directly or indirectly, and they also need basic knowledge of the hydrogeological concepts involved in aquifer development (Datta 2016). There exist obsolete paradigms about the origin, movement, and potential for pollution of groundwater. In any stressed aquifer, in order to address these issues it is essential to organize various types of educational activities aimed at different groups, from school students and teachers to officials of water administrations.

9.2.4.1 Expert Opinion

Comments by experts, analysts, and journalists in the media often sidestep objective assessment and are phrased carefully to avoid offending the government. To the public, these comments appear not sufficiently critical and independent of officialdom. Therefore, much expert opinion is liable to be compromised and watered down to the point of being not identifiable as criticism. The experts have a tendency to stay on the right side of governance because they worry about losing access. Hence, comments and views tend to mirror what the officials are likely to say, before policy decisions are taken. After policy decisions are made, all efforts are made to endorse them. As a result, governance does not really benefit from the views of the public or the experts. When the water resources ministry and the policy-making committee make an assessment of the water situation, it should be through a process, based on inputs from experts in academia, who have the actual field knowledge and experience.

The media usually sensationalize the issues, debating falsehood and facts, their own role, and possible ways to react. Most debate centers on how the new political order threatens scientific knowledge and research funding, or downgrades policy, but many overlook how science is losing its relevance as a source of truth. The realities about socially relevant science can be made clear if communicators, scientists, institutions, and funding agencies work together to present socially relevant science to the public, by encouraging changes in the content of what is being communicated and targeting various points of contact between science and the public. If the public were better equipped to navigate this, it would help to restore trust and improve understanding of different views, and perhaps help people to see through some of the fake news that circulates on scientific matters.

For example, while climate change may increase or decrease the supply of water, depending on a complex mix of rainfall trends and the greater evaporation caused by warmer temperatures, the actual impact on available water is hard to predict, and projections of impacts on the hydrologica cycle for a single region are often diverse. It is difficult to find concrete evidence of climate change having a greater impact than population growth on water availability. Scientists, experts, and journalists should try to explore the power of population growth in placing demands on finite supplies of fresh water rather than reflexively pointing the finger at climate change. While neither population growth nor climate change is amenable to easy solutions, there could be logical ways to address each, compatible with the values of human rights and reducing inequality.

In view of the above facts, it is desirable to motivate all stakeholders to stop giving credence to unreliable ideas with a sketchy scientific basis on water resources but rather to change behaviors, mindsets, attitudes, and practices to prevent/resolve conflicts (Datta 2014, 2015b, 2016, 2017). The media would serve the public better if much of the good work done so far by water resources professionals were made available to them so that aspects such as water rights, utilities, water resource assessment and societal use, which are of paramount importance to water governance and management, are best financed and conducted.

9.2.5 Behavioral Change Related to Water Rights

Self-interest and expected benefits usually influence public views on rights on groundwater. However, the National Water Policy, framed in 1987, revised in 2002, and debated as a draft in 2012, seems to be somewhat indifferent on a commitment to "right to water" as distinct from "water rights." While rights to divert water from water bodies and the water processing and delivery plants are profitable commodities, as water demand outstrips supply, considerable capital appreciation of these commodities cannot be ruled out. Governance has to ensure the equitability of water markets along with managing water demand, to avoid the possible buying and selling of water rights by wealthy water speculators and big corporations, leaving poorer people without water.

9.2.6 Behavioral Change to Manage Water Demand

Although agriculture is the largest user of water, links between agricultural land use and groundwater use have long been recognized, but these have not been widely translated into water demand/use management policy or practice. Self-interest usually influences the demands of the public on groundwater and land. Knowledge of the aquifer systems and groundwater demand/use in the social and economic context is relatively small and often limited at the level at which a management response is required (Datta 2015b, 2016). A significant part of non-revenue water is lost through broken irrigation channels and leaking pipes in the water supply distribution and waste-water treatment systems. In some states in the past, for better economic return, irrigation water demand increased due to growth in cropping intensity and replacement of crops consuming less water with more water-intensive crops (CGWB 1998).

Water demand can be decreased by checking water loss from broken irrigation channels, fixing leaking pipes, and reusing treated water, wherever feasible, by investment in technology for low-cost pipe materials in household water storage and sanitation, so that the cost of water saving is less than the cost of getting more water. In some cases, unbilled/unpaid for/stolen water is a bigger problem than the physical losses (Datta 2014). Increasing demand-induced groundwater vulnerability can also be minimized by strict enforcement of regulatory measures for water allocation, for short- and long-term restriction on indiscriminate groundwater abstraction, unplanned agro-chemical application and waste disposal in potential recharge zones (Datta 2015b, 2016, 2017).

9.2.7 Behavioral Change Related to Water Allocation

Distribution of renewable resources of available groundwater or fossil groundwater among potential or actual users may be a source of conflict between persons, institutions, or regions. Each case may be different according to the cultural, political, and legal background of the region. Nevertheless, it may be useful to try to achieve some kind of agreement on the ethical principles that should govern water distribution and management. Agriculture being the largest user of water, scope exists to prioritize water allocation by reliably quantifying the costs and benefits for agriculture and all other sectors, by decreasing agricultural water demand, improving productivity, encouraging the production of water-intensive and less water-intensive crops in areas with different comparative advantages, by increase in rain-fed crop yields and productivity, and irrigation efficiency at affordable cost (Datta 2005, 2015b, 2016, 2017). These goals may be supported by more effective approaches such as the following.

- Categorize the actual water needs (i) where the ratio of consumptive use to demand is high (e.g. irrigation) and (ii) where the ratio is almost negligible (practically all other water users).
- Conduct studies on the relevance/validity of competition among water users (private and public) and intersectorial use (irrigated agriculture and urban water supplies).
- Encourage non-sensitive groundwater users to switch from exploitation of high-quality aquifers to lower quality groundwater.
- Restrict abstraction rights for industries that have not installed water-efficient technologies by appropriate pricing for extraction.
- Change agricultural subsidies to encourage water conservation, improve water use efficiency, support community-scale projects, etc.
- Restrict or eliminate waste disposal and unplanned agro-chemical application in potential groundwater recharge and protection zones, identifying the sources of pollution and promoting containment of pollution from known sources.
- Provide incentives to use saline and brackish waters, and assess the possibilities to conserve floodwater in aquifers under river floodplains, wherever feasible.
- Assess past successes and failures and adjust policies according to local conditions. Examine people's adaptive strategies during water scarcity.

Considering these points, new guidelines and policies have to be developed on improved integrated and cost-effective basin-wide water resources assessment, development and management, based on holistic scientific knowledge and information on the aggregate situation of groundwater system, quantity and quality of water, land use changes, urban growth, pertinent parameters, possible changes in the hydrological cycle, and socioeconomic conditions (Datta 2005, 2013a, 2015b, 2017), in order to protect aquifers and drinking water supplies by alternative transparent institutional arrangements along with effective control of existing institutions that enable efficient water supply and distribution.

From the viewpoint of water quality, nationwide analyses should include evaluation of observed total dissolved solids (TDS) concentrations in groundwater, the three-dimensional probability of occurrence of brackish groundwater, and the geochemical characteristics of saline (greater than or equal to $1000\,mg\,L^{-1}$ of dissolved solids) groundwater resources. Aquifer-scale analyses may focus primarily on regions that have the largest amounts of observed brackish groundwater from previously published work; the distribution of dissolved solids concentrations; considerations for developing brackish groundwater resources, including a summary of other chemical characteristics that may limit the use of brackish groundwater and the ability of sampled wells producing brackish groundwater to yield useful amounts of water.

Waste-water quality may be better than available water sources in some cases or may contain more TDS, organics, or microbial agents than available surface and groundwater. In characterizing the water or waste-water source, every aspect over time is critical. Some of the parameters which should be routinely monitored include pH, residual chlorine, conductivity, phosphate, ammonia, chloride, sulfate, total iron, manganese, silica, biological oxygen demand, chemical oxygen demand, and scaling index. Some of the chemical constituents may cause the following problems.

- Mineral scaling from calcium carbonate, calcium phosphate, and other products.
- Corrosion, pitting, and stress damage to metal heat transfer surfaces and to structural metal surfaces (e.g. damage to copper, copper alloys, and other "yellow metals" from ammonia).

It is critical to do this to provide design engineers and chemists with an accurate picture of the chemistry and water flow patterns in order to:

- determine if additional treatment processes are required
- properly design any required treatment system
- determine if treatment is needed to avoid scaling or corrosion
- develop necessary treatment approaches
- determine if storage capacity is required to compensate for flow variations.

9.2.7.1 Co-ordination of Elites and Non-elites

Humans are naturally egotistic, competitive, and nationalistic, resulting in polarization among non-elites and their submission to elites. Since most people compete for jobs, opportunities, social space, and other means to "make it in life," there is little unity and resistance to elite rule. Hence, people are submissive to elites, accepting their perceived superiority and power as outcomes of individual merit, resulting in greater empowerment of elites and leading to power imbalances in societies. The elite are seldom collective or homogeneous, but impose their preferences using their position and opportunity.

Those who control the means of interpretation and communication diffuse their own experience and culture as the norm. The irony is that political semi-elites often share similar values, ideologies, and interests with the economic elites, but are lower in the power hierarchy. It is the managerial quasi-elites such as academia, polity, bureaucracy, staff of international organizations, transnational corporation managers, and other upper middle-class individuals who do the thinking and drafting according to the directions of the higher powers.

More research is needed to enhance knowledge in fast, efficient and effective diagnosis in complex environmental matrices for rational approaches to identification of source water pretreatment needs and acceptable uses of recovered water. Clear and integrated information is required regarding impacts and policy options for water scarcity mitigation and adaptation at state and national levels, and about their respective costs, benefits, and risks which can better inform decision makers and societies about the consequences associated with alternative choices.

Governance along with researchers need to execute their roles in recognizing and managing diverse viewpoints, without compromising scientific rigor and objectivity. Operating single teams and technical support units from multiple geographical locations should be avoided, to ensure that management tasks do not get further complicated and the high costs of time and travel commitments do not preclude the participation of the best researchers (Datta 2015a, 2016, 2017). For effectiveness, scientifically independent organizations, not driven by political or economic self-interest, should do the baseline work in order to ensure an unbiased result.

To reduce water consumption and wasteful utilization, practical measures could include strict regulatory enforcement of groundwater extraction; prevention of waste discharge into the hydrological system; identification of pollution sources and strategies for containment of pollution spreading from known sources; development of maps of groundwater vulnerability; conservation of floodwater in aquifers under river floodplains; and delineation of potential groundwater recharge and protection zones. It is necessary to assess past successes and failures and adjust policies according to local conditions, conduct studies on competition among water users (private and public) intersectorial use (irrigated agriculture and urban water supplies), and examine people's adaptive strategies during water scarcity.

Future demand for fresh water could be reduced by fixing leaking pipes in the water distribution system, reuse of treated water, and saline water agriculture, wherever feasible, investment in household sanitation, water storage, treatment of industrial effluents in multipurpose water schemes, and construction of eco-friendly pipelines and aqueducts to move water from areas of abundance to areas of scarcity (Datta 2014, 2016). Water crises might occur unless legislation is updated for more reliable, transparent, and consistent integrated water management in policy making, such as cost-effective improved basin-wide management to protect aquifers and drinking water supplies, and efficient institutions that enable water to be supplied and distributed expeditiously. The subsidies in agricultural water should be changed to encourage rainwater collection, water conservation, producing more food with less water, and selective introduction of community-scale projects.

While safeguarding groundwater is a challenge, most solutions are found at the aquifer, watershed, or local level. An external force is often required to achieve necessary changes and accountability, but top-down management usually sets up resistance among stakeholders. There is virtually no possibility of getting entrenched groundwater users on board if they aren't actively involved in the decision-making process. Unfortunately, political and financial support for even the most basic data collection is usually a low priority. For urban/rural water supply planning, there is a need for a systematic evaluation of the aggregate status/situation and use of surface water and groundwater to meet future demand.

9.3 Ethical Issues in Groundwater Governance and Use

Although ethical perceptions may vary from person to person and among societies and countries, for the protection of groundwater from depletion and degradation, in order to achieve sustainable water supply, this vital resource has to be developed and managed, taking into consideration the environmental as well as social, economic, geographical, and political aspects. Additionally, the actions and decisions of an individual or group have to be governed, and practical and social decisions have to be treated ethically. The welfare and safety of citizens and economic development should be given more importance than unscrupulous practices working hand in hand with greed and private gain (Datta 2005, 2013a, 2015b, 2017). This would require a change in attitude and behavior as well as enforcement of strong anticorruption rules and regulations.

In this perspective, "groundwater ethics" can help to increase public awareness of the interconnection between natural and human systems and interactions between people, their habits, their culture, and the natural landscape. To manage, prevent or mitigate unexpected hazardous consequences, we need to educate all stakeholders (including politicians, policymakers, media, and public) to consider "groundwater ethics" in their choices, and how they can overcome impacts by critical analysis of ethical dilemmas, guided by the best obtainable detailed scientific information and socioeconomic considerations (Datta 2013a, 2015b, 2017).

To ensure proper selection of sites for the construction and maintenance of RWH groundwater infrastructures, and to characterize spatiotemporal variability of groundwater regimes, an integrated disciplinary approach by reliable, competent, capable and transparent institutions, avoiding corrupt and unscrupulous practices, can be helpful. For better governance, strategies, management plans and policies, involvement of corrupt and unscrupulous practices for private gain should be discouraged, to enhance efficiency and promote harmony through trust, knowledge, transparency, respect of interests, and sharing responsibility/resources (Datta 2013a, 2015b, 2017). The following ethical issues may prove relevant to achieving sustainable water use.

- *Control neglect of groundwater problems*: the main cause of the pervasive neglect of groundwater problems among water managers and decision makers is probably the perverse hidden or open subsidies that have traditionally been a part of large surface water irrigation projects. Surface water for irrigation is usually given almost free to the farmers and its wasteful use has become the general rule. Hence, more comprehensive attention during groundwater planning and management would be necessary to control this.
- *Declaration of groundwater as common resource*: for sustainable groundwater management, the legal declaration of groundwater as a common resource is essential to promote solidarity in its use. Nevertheless, management of groundwater is a big challenge in many countries despite groundwater being in the public domain for many decades. To overcome this challenge, stakeholders should be made responsible for groundwater management under relevant water authority guidance.
- *Transparency in sharing groundwater data*: adequate hydrogeological knowledge and reliable information and education are crucial to facilitate co-operation among aquifer stakeholders. The traditional attitude of water agencies of not allowing the public easy access to water data needs to be changed, and making use of information technology, all stakeholders should be given easy access to reliable data on abstractions, water quality, and water levels.
- *Ethics of pumping non-renewable groundwater resources*: in regions such as the arid Rajasthan Desert, with very small amounts of renewable water resources but significant amounts of fresh groundwater reserves, groundwater mining may be a

reasonable action if various conditions are met: (i) the amount of groundwater reserves can be estimated with reliable accuracy; (ii) the rate of reserve depletion can be guaranteed for a long period (e.g. 50–100 years); (iii) the environmental impacts of such groundwater withdrawals are properly assessed and considered less significant than the socioeconomic benefits from groundwater mining; and (iv) solutions are considered for the time when the groundwater is fully depleted.

The current thinking in the water sector is that strategies need to be developed to ensure utilization of groundwater resources within the renewal capacity. However, it is recognized that quantification of sustainable use levels requires extensive research. The frequently expressed view that the water policy of arid zones should be developed in relation to renewable water resources is unrealistic. The ethics of long-term water resources sustainability in arid regions must be considered with careful management to utilize resources beyond the foreseeable future.

9.4 Role of Geoscientists

Professional geohydrologists should transfer their awareness of the uncertainties to decision makers and the general public, with prudence and honesty, in order to avoid loss of credibility of the scientific community, either in the short term (by giving the impression of lack of knowledge) or in the medium term (due to the failure of predictions to be realized). The "gloom and doom" pessimistic predictions made by certain individuals and institutions about the depletion of natural resources or the population explosion have usually failed to materialize.

In any riverine system, when sudden rainfall-induced surplus runoff occurs, for runoff resilient groundwater management to sustain water supply, generally knowledge on the geohydrological system and landform heterogeneity is limited and insufficient (Datta 2013a, 2015b). Geoscientists with wider experience can help to map and analyze changes in topography, land contours, stream and river flows, flow paths, river beds and flood plains, land classifications, and hydrological calculations, and ensure proper planning and preparedness strategies, in order to manage groundwater resources.

They can also involve school children in programs at local level to educate and create awareness through the latest technical and scientific advancements, new communication strategies, and by interaction with politicians, media, and citizens (Datta 2013a, 2015b, 2017). This can be helpful in promoting their ethical, cultural, socioeconomic, and educational values, and changing mindsets, strengthening relationships and framing policy guidelines on water conservation, to develop a problem-solving, holistic interdisciplinary approach, planned in association with the public, by critical analysis of geo-ethical dilemmas.

Partnerships between geoscientists and communities may help to generate new resources and more efficient methods, especially in floodwater management and conservation. The successful design/operation of a groundwater abstraction and pollution monitoring and surveillance system should be based on determining the hydrogeological characteristics of the groundwater flow field under natural and stressed conditions, dynamics of water pollutants, and linkage with spatial and temporal variability in concentration.

Iso-concentration maps of contaminant levels should be prepared for all regions and revised from time to time, in relation to changes in land use patterns. To protect the resource base and enhance its efficiency, the focus of research should shift to eco-regional aspects, realizing the significance of close interlinkages of water resources, environment, physical sciences, earth and atmospheric sciences, etc. Investigations must also cross national/state borders by mapping and quantifying fresh groundwater, and creating a common strategy for its exploitation.

9.5 Role of Media

Although there have been instances where the media contributes to pushing for more transparency and accountability, these cases are still rare. Overall, coverage on issues of corruption and lack of integrity is low, which is an issue in many countries. It has been argued that stories of corruption in the water sector fail to get the necessary coverage in the mainstream media compared with issues such as business and politics. Coverage of corruption or lack of integrity in the water sector is often biased and predictable, which is a more serious issue. The lack of coverage and at times biased coverage on corruption or lack of integrity shows that there is still a way to go before the media plays its potential role of encouraging and catalyzing change within the water sector. In most countries, the media does not know how to tell the story of corruption in the water sector in a compelling enough way to grab the public's attention. Since this is a sensitive topic, there are limitations to how the media can position itself effectively in advocating less corruption in the water sector.

The media, along with other agencies, can play an important role in corruption detection and promoting transparency and accountability in the water sector. The news media can help to raise public awareness of corruption through investigative journalism, news reports, dissemination of research findings, and so on. The government, private sector, and/or citizens can then agitate for action in the form of launching an investigation or judicial action, or calling for the dismissal or resignation of corrupt officials, thereby contributing to improved transparency and accountability. In order to fight corruption in the water sector, there is a need for people to first recognize that corrupt practices exist. Local and national media both have an important role to play in bringing issues of corruption to the attention of the public and policy makers, to ensure that action is taken through policy or advocacy.

Several things come into play here. First, ownership of the media can play a role. The question of whether the media is independent or state-owned influences the extent to which it can be critical about the level of corruption in state institutions. State media tends to be less critical of government institutions, whilst the private media will most likely be more critical. Ownership of the media to a large extent determines how robust and critical it can be in fighting corruption, since media owners in some cases exercise considerable influence over its editorial stance. Government-owned media is designed to a large extent to represent the state, and to portray that state and its agencies in a favorable light. Furthermore, the amount of resources available to journalists may influence how effectively the media is able to act as a watchdog in fighting corruption.

Nearly every day, both the state and private media are invited by government agencies and other organizations to cover events of some sort. After these events, the journalists are provided with some form of compensation – usually known as "solidarity." It is also very common for some government officials to chauffeur these reporters to and from their offices as a strategic way of maintaining patronage relations and a mutual relationship of need and obligation. This therefore limits a journalist's ability to be critical. It has been argued that such practices equate to corruption in the media itself. Journalism, like any occupation that relies heavily on personal integrity, demands a high degree of sacrifice but pays off in the end, when one's story is published or one receives a prize for one's work. Integrity issues in this line of work become even more apparent when journalists themselves do not realize the issue by accepting bribe.

The rise of the internet and the ease of posting articles online have made it difficult to assess which reports are accurate or reliable. Social media presents opportunities as well as challenges for the future of the news media in promoting integrity in the water sector. It offers many people new ways of networking, and of sharing and receiving information outside the mainstream media such as TV, radio, and newspapers. The credibility of social media depends on the media house and the reporter – some media houses go to people's personal social media pages for news. Als,o there is false news on social media as there is a fight for attention; as such, there is skepticism about information on social media. A growing public perception exists that journalists are irresponsible, unfair, biased, and, above all, unethical. There are many reasons for this; the possibility that journalists have abandoned their ethics and objectivity for political partisanship is one of them.

The watchdog role of the media is not limited to producing information about misbehavior, but also concerns how that information is used to hold people accountable for their actions. A government must know that people want responsiveness and wish to hold those in power accountable for their actions. A country's media is likely to have a minimal effect on corruption if it toes the political line or fails to obtain the necessary support from the government, the private sector, and civil society. It is important that new initiatives are established where the media is further encouraged to take a keen interest in reporting on water-related issues. International non-profit organizations, as well as other civil society organizations, have a role to play in ensuring that journalist networks are supported to report on these issues. It is important that the interest of journalists in reporting on such issues is sustained, which could be done through involving them in training courses or broadening their knowledge and awareness on integrity issues in the sector. The government has a role to play in ensuring that the space for the media remains open and that their ability to report on sensitive issues is assured.

All organizations should intensify their efforts in supporting the media to report on water issues. Journalists who show an interest in the water sector should be given the opportunity, through training courses, to broaden their knowledge and awareness of integrity issues. There is a need for enhanced monitoring mechanisms to be utilized by citizens, civil society and the media in order to strengthen accountability and transparency, and to ensure value for money in water service delivery.

9.6 Guidelines for Rainwater Harvesting Projects

9.6.1 Assessment of Local Rainfall

There needs to be an assessment of total annual rainfall, monthly rainfall, maximum hourly rainfall intensity, and variations in rainfall over a period of years. The rainfall data for the nearest station can be accessed from published literature and on websites of concerned organizations.

9.6.2 Site Plan of the Premise

Create a detailed site plan with scale, demarcation of built-up area, paved area, park/fallow land area, storm water and sewerage drain lines, existing functional and non-functional groundwater abstraction structures, septic tank or waste disposal site.

9.6.3 Availability of Surface Runoff

Estimate the total available hourly surface runoff for the maximum intensity rainfall, separately for the different land uses, and then sum to give the total maximum hourly surface runoff available. An estimate of total monsoon and non-monsoon runoff should also be made separately, which can be added to give total annual runoff available. Any committed use of the surface runoff also needs to be studied.

9.6.4 Assessment of Depth to Water Level and its Long-Term Variations

The existing depth to water level needs to be assessed. Sometimes the site may not even have an operational well, in which case the water level data of the nearest observation well monitored by the nodal organization can be used for the purpose. The temporal long-term pre- and postmonsoon variation in water level of the locality also needs to be assessed.

9.6.5 Groundwater Salinity

It is always useful to have an idea about quality of groundwater at the site. The TDS manifested as salinity gives a first approximation of groundwater quality. This can easily be done by measuring electrical conductivity (EC) or TDS of the groundwater with a hand-held pen-type EC or TDS meter.

9.6.6 Local Subsurface Geology and Aquifer Disposition

The subsurface geology indicating lithological variations with depth and demarcation of the aquifer are important items of information. Thus, a local subsurface fence diagram or cross-section is required. This can be made from lithologs obtained during any drilling operation or from data available in literature. If no data are available then a resistivity survey needs to be done.

9.6.7 Decisions Regarding Further Course of Action

Before recommending artificial recharge (AR) to groundwater:

- ensure sufficient availability of non-committed surplus rainfall runoff. If there is sufficient rainfall runoff available, RWH by means of storing all or part of the runoff for lean period use can be attempted
- assess whether the groundwater level scenario at the site permits AR. Ensure that the site normally has a sufficiently deep postmonsoon water level so that further rise in groundwater level by AR does not create water logging and soil salinity, and the postmonsoon water level rise after the AR project should always have over 3 meters of

vadose zone. For implementation of AR, the local groundwater level should have a temporal declining trend

- in areas with highly saline and shallow groundwater with no significant decline, storing the runoff for lean period use can be planned for RWH
- in areas with highly saline or marginally saline relatively deeper groundwater, and significant declining trend, combining runoff storage and AR structures can be planned
- when AR projects are to be implemented, the local subsurface geology and the aquifer disposition help in selection of the type of recharge structure and its design parameters.

References

Central Ground Water Board (CGWB) (1998). *Report on Groundwater Resources of India*. New Delhi: Government of India.

Datta, P.S. (2005). Groundwater ethics for its sustainability. *Current Science* 89 (5): 1–6.

Datta P.S. (2013a). Groundwater vulnerability to changes in land use and society in India. Understanding Freshwater Quality Problems in a Changing World, Proceedings of H04, IAHS-IAPSO-IASPEI Assembly, Gothenburg, Sweden, July, pp. 345–352.

Datta P.S. (2013b) Geoethics and geoscientists role to protect geodiversity for societal development. GEOITALIA2013, Session Geopolicy – L 2 Geoethics and Society: Geosciences Serving the Public, September 16–18.

Datta P.S. (2014). Need for better groundwater governance for sustained water supply. IPHE National Conference on Piped Water Supply and Sewerage Systems, January 24–25, New Delhi.

Datta P.S. (2015a). Traditional rainwater harvesting at crossroads – time to seek ethical opportunities for reform. Proceedings of Aqua Foundation IX World Aqua International Congress, "Reviving Traditional Water and Environment Conservation Techniques," November 26–27, New Delhi, pp. 93–115.

Datta, P.S. (2015b). Ethics to protect groundwater from depletion in India. In: *Geoethics: The Role and Responsibility of Geoscientists* (ed. S. Peppoloni and G. di Capua), 19–24. London: Geological Society.

Datta P.S. (2016). Ensure smart water availability, by behavioral change. 10th World Aqua Congress, "Water- Smart Solutions for Growing India," November 24–25.

Datta, P.S. (2017). *Better Groundwater Governance Only Can Ensure Sustained Water Supply*. Germany: Lambert Academic Publishing.

Further Reading

Conti, K.I. and Gupta, J. (2015). Global governance principles for the sustainable development of groundwater resources. *International Environmental Agreements* 16 (6): 849–871.

Datta, P.S. (2008). *Water: A Key Driving Force*. New Delhi: Vigyan Prasar Publications.

FAO (2015) Shared global vision for Groundwater Governance 2030 and A call-for-action. Available at: www.fao.org/3/a-i5508e.pdf.

10

Concluding Remarks

CHAPTER MENU

10.1 Scope of Rainwater Harvesting

Water resource characteristics and distribution differ from region to region. Many water-scarce regions have water demands which far exceed supplies, with subsequent vulnerability to hydrological stresses, such that they require exogenous water. Enhancement in safe water availability and supply will be determined by the policies, plans, and technologies available, in addition to political, socioeconomic, and other factors. For effective implementation of water demand and supply management strategies at local and regional level, better governance requires that each area/region be investigated separately, by systematic monitoring, research on changing patterns of water use from different sources, and dynamics of pollutants in groundwater flow under natural and exploited conditions. To effectively manage water demand and wasteful use, development and investments efforts have to be approached holistically, by competent, capable, and transparent institutions, taking into consideration environmental and social returns, as well as financial returns.

In most water-scarce regions of the world, including India, despite successful cases of rainwater harvesting (RWH) at local level, RWH offers extremely limited potential to reduce the imbalances in water demand and supply, and provide reliable supplies in water-scarce regions, because significant areas of such regions are characterized by low mean annual rainfalls, high interannual variability in rainfall, and high potential evaporation (PE). A large part of such variability occurs during the rainy season, reducing the runoff potential and increasing the occurrence of hydrological stresses. In regions which have medium rainfall but experience medium to high evaporation, the poor groundwater potential, which underlies these regions, poses a constraint for recharge. The overall potential of RWH would be high in regions with high rainfall and medium evaporation. For instance, in India, a large part of the water-scarce regions, which fall

Water Harvesting for Groundwater Management: Issues, Perspectives, Scope, and Challenges, First Edition. Partha Sarathi Datta.

under the medium rainfall and medium to high evaporation regime, are underlain by hard-rock formations such as basalt, crystalline rocks, and other consolidated formations such as sandstones. In such hard-rock areas and also in areas with silty clay and clayey soils, percolation tanks are likely to have low efficiency. In these regions, planning of recharge schemes should consider surface water impoundment of all the available excess flows, rather than direct recharge.

10.2 Economic Evaluation of Water Harvesting Systems

Such evaluations pose several complexities due to problems in quantifying the hydrological impacts and the various benefits. Due to difficulty in quantifying the inflows, storage and recharge efficiency, and the economic value of the incremental benefits, economic evaluation of costs and benefits of RWH and groundwater recharge systems poses several social, direct economic and ecological or environmental complexities. The economics of water harvesting cannot be worked out for structures on the basis of individual benefits and costs when the basin has limited surplus water but rather should be done on the basis of incremental benefit. In many water-scarce basins, there is a strong tradeoff between maximizing the hydrological benefits from RWH and making them cost-effective. In many water-scarce basins, RWH interventions lead to distribution of hydrological benefits, rather than augmentation. There is an optimum level of water harvesting which a basin can undergo to help optimize the gross value product of water in relation to economic, social, and environmental outputs basin-wide. The higher the degree of basin development, the higher would be the marginal cost and the lower the marginal benefit from water harvesting.

In basins that experience high interannual variability in stream flows and/or groundwater level, the tradeoff between the hydrological impacts of water harvesting and economic benefits is likely to be large. With increasing storage capacity of RWH systems, the economic viability decreases as the average cost of water harvesting per unit volume of water increases. In closed basins, there is apparent tradeoff between local benefits and downstream benefits. Upstream diversions reduce the prospects of storage and diversion systems downstream. In many basins, there is an apparent tradeoff between maximizing overall benefits for communities in terms of enhancing the gross value product of water, and maximizing the local benefits of water harvesting in upper catchments. This is because in these basins, water from well-endowed regions with low water demands is being diverted to poorly endowed regions with high water demands, enhancing its social and economic value.

10.3 Scale Considerations

These aspects are also extremely important in evaluating the cost and economics of water harvesting. Sometimes, potential social benefits of improved regional equity may influence political decisions to intensify upstream water harvesting, even if it may reduce net benefits at the basin level. In developing countries, including India, RWH efforts made so far have seldom included spatial and temporal scale considerations when analyzing the physical and economic impacts of water harvesting. These aspects are important almost everywhere where high interannual and interseasonal variability in rainfall exists. Issues for concern include (i) optimal level of water harvesting in different river basins that averts unintended downstream impacts; and (ii) ecological and environmental impacts of water harvesting in terms of reduction in environmental flows, or increase in lean season flows in different hydro ecologies.

10.4 Advantages and Disadvantages of RWH for Groundwater Recharge

10.4.1 Advantages of RWH for Groundwater Recharge

- Subsurface water storage space is available free of cost.
- Water evaporation losses are negligible.
- Water quality is improved by infiltration. Biological purity is very high.
- No adverse social impacts occur, such as displacement of population, loss of scarce agricultural land, etc.
- Water temperature variations are minimized.
- Environment friendly, controls soil erosion and flood.
- Water stored underground is relatively immune to catastrophes.
- Provides a natural distribution system between recharge and discharge points.
- Possibilities for energy saving due to reduction in suction and delivery head as if there is enough rise in groundwater levels.

10.4.2 Disadvantages of RWH for Groundwater Recharge

- Additional cost of pumping the groundwater.
- Complex operation.
- Degradation from pollution sources can be long-lasting if not managed carefully.
- Institutional, organizational, and legal problems due to groundwater being a common resource.
- Requires periodic/premonsoon maintenance and involvement of user for monitoring.

10.5 Practical Suggestions for Efficient Water Harvesting

10.5.1 Enhancing Knowledge of Catchment Hydrology

In most efforts for RWH in developing countries, the catchment hydrology is not well understood. The focus of any rainwater-harvesting project, particularly in India, is collecting the excess surface runoff water, storing and subsequently diverting it for beneficial uses. Most small rivers are not assesssed for stream flows and siltation. Scientific data on hydrological parameters such as soil infiltration characteristics, weather patterns, land use characteristics, and catchment slopes are essential to arrive at reliable results. Extrapolating data for large reservoirs to small catchments can lead to erroneous estimation of runoff and siltation rates, as siltation rates are generally high for hilly upper catchments. Comprehensive evaluation of the severity and extent of surface runoff and groundwater pollution demands more studies. Basic scientific information on water quality at and around the RWH sites is scarce and the cost of acquiring the necessary new information is relatively high. The monitoring strategy needs to be changed from general state descriptions to problem-oriented information on groundwater renewal, distribution, and recycling.

10.5.2 Research Focus

It has to be kept in mind that green water in the soil profile, which is used directly by natural vegetation and crops, is also an important component of the hydrological system and there has been no focus on improving efficient utilization of the water harvested in tanks, khadins, percolation ponds, and johads. In high rainfall regions, the utilizable surface water resources are much less in comparison to the runoff generated. Here, effective strategies to capture runoff *in situ* for crop production through proper land use planning would help improve both surface and groundwater use, and alter the hydrology positively. In river basins which experience high aridity during the summer months, storage of water in tanks, ponds, and other small reservoirs can lead to heavy losses through evaporation.

For any water-scarce river basin, before planning any water harvesting and recharge project, the most important aspects are to know whether the basin has any surplus flows or natural sinks, or if a significant amount of water is lost in evaporation from natural depressions. Both water accounting and water balance studies, to examine what percentage of the water could be captured without causing negative effects on downstream uses, should be carried out for typical rainfall years so as to capture the hydrological variability. If this is prevented, it can lead to wet (or real) water saving, through increase in output per unit of depleted water. Directly diverting the harvested water from the RWH system to the agricultural land is critical for maximizing the net hydrological gain, especially in areas with poor groundwater storage or areas experiencing high interannual variability in runoff.

From the point of view of water quality at the RWH site, it is desirable to develop uniformity in the methods for selecting wells and more information for deeper wells. It is essential to monitor and assess temporal variability in water pollutants and links with accumulation and leaching characteristics of chemicals, groundwater dynamics, aquifer attenuation capacity for pollutants, pathways of pollutants between the source and the water supply well, under natural and exploited conditions, in relation to pattern of land/water use. This information will be helpful to delineate potential groundwater recharge zones and protection zones. In the RWH locations, any case of groundwater pollution suggests the beginning of a trend of increasing chemical residues in groundwater, in response to prolonged use of agro-chemicals and indiscriminate disposal of wastes. The scope of the problem may be larger than imagined and assessed so far.

Schemes are normally taken up in the following areas:

- where groundwater levels are declining over a period of time
- where substantial numbers of aquifers have already been desaturated with water
- where there is availability of surface water for recharge during the rainy season, and availability of groundwater is inadequate in lean months
- where salinity ingress is taking place, and where there is a quality problem in groundwater.

At different scales, the characteristics of pollutant levels and transport in groundwater are altogether different, and are associated with variations in one or two parameters at one scale and several parameters at another scale. A broad qualitative knowledge of most of the effects is available but little is known about the interactions between solute and soil matrix, which control the behavior of pollutants. Since most of the various parameters and physicochemical processes as a whole are not completely understood, it is difficult to describe the exact pattern of pollutant migration within natural boundaries. Moreover, RWH ponds/tanks may be connected to groundwater flow and may be subject to contamination by dissolved chemicals.

While there are still areas requiring research, from the point of view of action, the following steps seem to be important to make water harvesting more efficacious:

- developing a better understanding of catchment hydrology
- developing basin water accounting and balance
- focusing on wet water saving
- enhancing the productivity of green water in the basin.

Since the responsibilities for effective governance of water are divided between the central and state authorities, due to inadequate co-ordination among departments, implementers, and regulators, there are gaps in the overall vision to find sustainable solutions to water scarcity, and fragmented approaches are taken. Hence, implementation of the aforesaid approaches is likely to face several financial, technical, and legal difficulties, with respect to the jurisdiction of different agencies, and benefits from these approaches can be assessed only after the system has worked efficiently for a few years. Governance can provide better management by consolidating problem-solving approaches, with detailed insight into processes governing changes in land use and water demand.

In any area, groundwater flow often contains a transient component governed by natural groundwater recharge rates, discharge altitudes, and hydraulic properties of the aquifer system. Current groundwater withdrawals generally have both recent and past recharge components. The present to past recharge ratio is large in humid areas and small in arid areas, and transient conditions prevail at low recharge rates and are steady at high recharge rates. Hence, governance should provide better groundwater management by not ignoring the transient component in areas of low/high recharge and high/low discharge, so that groundwater safe yield is neither overestimated nor underestimated, arresting depletion and sustaining development in the long run.

To protect groundwater from further depletion, governance should introduce better legal provisions and strictly enforce them to restrict indiscriminate withdrawal of groundwater, with emphasis on enhancing water use efficiency. Groundwater extraction has to be better governed somehow by balancing the recharge estimated not only on the intensity and distribution of modern seasonal rainfall but also from improved direct methods; such estimates need to be revised from time to time (Datta 2015). Wherever feasible, governance should adopt approaches to enhance recharge through low-lying areas, where surface runoff is collected during high-intensity rainfall, instead of allowing the water to stagnate. To protect groundwater from degradation, recharge zones should be clearly identified and revised in relation to land use changes, and waste disposal activities in these areas should be restricted or eliminated (Datta 2013, 2015).

The solution to regional and local water problems requires education, technical assistance, and supporting research. It is imperative for better governance that the community at large participates in policy formulations and judgments of what is to be sustained. Strong public education and outreach programs are needed to improve understanding of the nature, complexity, and diversity of groundwater resources, and to emphasize how this understanding must form the basis for operating conditions and constraints. This is the only way to positively influence, for the long term, the attitudes of the various stakeholders involved. Pressure from the community for better management of groundwater resources will be the main driving force.

In reality, the aggregated results of individuals' selfish personal choices and their economic circumstances and social inequalities are always dependent upon the available alternatives, which vary greatly among different groups and individuals. In order to make the reality more bearable, people may "choose" to behave in a certain manner. Therefore, given that power is the most efficient when the least observable, governance should ensure that both the organization and accumulation of power take place in subtle ways, in complementarity with efforts to change behavior by raising stakeholder consciousness.

Although economic development in any country is highly dependent on a water supply of adequate quantity and quality, economic outcomes not only affect the economic sphere but also the human environment, behavior, and other non-economic realms. Governance should ensure that economic discourses, which are usually defined in material terms, do not only carry implicit positions and assumptions about the economy, but also think about humanity, the environment, and other matters that affect people. Often, economists, devoted to the maximization of utility and minimization of disutility, enjoy high prominence and the predominance of economics supports the interests, values, and ideologies of the already powerful and aspirants to power. Other related academic disciplines adopting its basic approaches also finance and strengthen economics. Its ethical tenets are also popularized through media, books, music, and other forms of cultural expression.

For decades, water resources professionals have been quietly working for the betterment of society but the role models are always some cinema actor or actress or a cricketer or a celebrity. The media also pays them large amounts of money to appear in advertisements, even if they have zero knowledge about water and environment issues. Unfortunately, in this way, the public is being misinformed and it appears as if only celebrities and some environmental activists are concerned about India's development, despite the very best professional brains working hard to improve matters.

The media would help the environment more if the achievements of water resource professionals could be projected properly! Most such professionals toil without recognition but in order to grow, become relevant and meaningful to changing times, any knowledge needs scrutiny and criticism, scholarship, rigor, original ways of thinking, and intellectual honesty. Metrics need to be developed for evaluating what might work, and how money can work for the benefit of all, leaving aside greed about multiplying its value. It is desirable for governance to learn and comprehend the knowledge and apply it to modern society in a justified manner, with a shift in focus on funding priorities from technological solutions to understanding, predicting, and changing human behavior.

The RWH evaluation studies reported so far are highly qualitative, mostly with respect to change in water availability, increased groundwater table, revival of flow of the rivulets, and yields of crops, based on secondary information collected from the farmers/beneficiaries; even the funding agencies experience limitations in implementing this at local scale. Geo-scientists have a significant role to create public awareness about the revival of RWH approaches. The well-known signs of groundwater depletion and erratic rainfall are forcing local communities as well as governments to consider local water harvesting and recharge structures on a massive scale, to increase groundwater availability, and to drought-proof and protect livelihoods. With a long history of artificial recharge in India, a variety of methods are available to further expand this technology. However, the popularity of artificial recharge is a function of its local success and the potential of numerous such local operations on basin-scale water availability and distribution, the techno-economic feasibility of various approaches, and cost and benefits are less well understood.

Moreover, "successful" local AR efforts can cause further problems. The cost mainly depends on the distance over which the source water has to be transported, stability of structures and resistance to siltation and/or clogging, and the degree of treatment of the source water. In general, the initial investment and operating costs of the recharge structures (especially in water-scarce areas) are less than those required for supplying potable water using tankers. Moreover, there is no evidence in the archival records that small earthen bunds were traditionally the preferred mode of RWH and irrigation, or that they provided an effective solution to water crises in periods of drought. However, systematic R&D programs on physical, economic, environmental, and institutional aspects and an improved understanding of the performance of such systems are required for better siting and effectiveness, which can resolve problems across scales to shape a more meaningful future.

Dealing with the basic issues, challenges, and limitations of water security such as climate forcing; rapid growth in population, urbanization, agricultural intensification, economic aspirations, groundwater overabstraction, and other mega-trends; SWOT analysis of water harvesting systems for groundwater management; overcoming knowledge gaps; and a demand management framework can guide communities towards achieving water security and consumption targets in a sustainable manner, by modifying the attitudes and behavior of all water users for the purpose of balancing demand with supply. This is likely to help the water industry to develop innovative cost-effective approaches and technologies for groundwater protection from depletion and degradation, and avoid costly economic, environmental, and political expansions to the water supply network.

The discussions in the previous chapters are important because better tools and approaches are needed to understand how a changed environment will shape our future world, and different measures are also needed to make decisions to allocate resources towards adaptation or mitigation. This can be achieved only if better information about costs and consequences is available. The concepts of water harvesting and water security, and the various demand management strategies and tools available can help water managers to modify the attitudes and culture of individuals and society to achieve sustainable water consumption targets. The case studies and the risks associated with groundwater overabstraction and managed aquifer recharge can be fruitfully used to achieve conservation targets to achieve water security.

Many developing countries are also suffering from a shortage of skilled water engineers although the manpower for operation and management is not a scarce resource. The argument of companies that engineering, procurement, and construction (EPC) teams can manage the operations is totally unjustified. Water sector operation and maintenance is traditionally handled by government agencies so most of the experienced workers have a government background. However, most government staff have little knowledge about the latest technologies, which poses challenges in managing the water supply. Until the sector is reformed and provides high returns on investment, it will not be able to attract good talent.

Water resources are becoming more stressed and water demand continues to increase. This is forcing the water industry and regulators to evaluate approaches and technologies and alternative sources of water for non-potable use. In some developed countries, treated municipal waste water is abundant and is a viable option for agricultural purposes, as cooling water, boiler feed water makeup, and cooling pond makeup. Most of the up and coming managers in the water sector are either good human resources managers or good program/project managers, who rely on the people they manage to have the technical/operational knowledge they do not have. Numerical dexterity in managing data is not enough. Investigating the resilience of systems/operational networks requires an understanding of the aggregate situation by competent and highly skilled people with sound technical knowledge, and practical, longer-term experience. It is generally felt that such people are not always appreciated. The water industry needs more than just engineers. Having qualified workers is essential.

10.5.3 Components of the Plan

- Tanks can be set up at parks, urban farms, and urban gardens for irrigation and for businesses to supply non-potable water.
- The utilities could set up rain tanks and sell water to businesses.
- Selling rainwater to businesses will generate more revenue than domestic water.
- Storm-water management is necessary to reduce flooding and stream pollution.
- Training and testing facilities should be available for city inspectors, designers, and contractors.

10.5.4 Advanced Solutions to Water Security Problems

- Better integration and co-ordination of water governance, planning and services, by creating a national capability to forecast floods, droughts, water quality, and water supply.
- Reduction of flood damage through more active and integrated river basin water management, assessing flood risk and restricting development in future flood zones.
- Reassessment of infrastructure, and capability to manage and store water, in expectation of droughts longer and more severe than any experienced.
- Management of the cumulative effects of development in watersheds to reduce contamination of lakes and rivers, so that the water is safe to drink and sustains aquatic ecosystems.

Managing rainwater and storm water efficiently would ensure plenty of water for future generations. To achieve these water security goals, more co-ordinated, inclusive, and effective water governance is needed to measure, predict, and manage water, instead of being fragmented between several federal and provincial ministries. Indigenous people are left out of water governance and policy decisions, despite their rights and the high exposure of many indigenous communities to drought, flooding, and impaired source water quality. Advances in science, prediction, measurement, and policy analysis can contribute to avoiding water disasters.

Expanding the surface water reservoir system, with the cost of land, new infrastructure, and energy use, seems futile in the face of water loss due to evaporation, runoff linked to groundwater pollution and other environmental issues. Water companies can evaluate the financial value of the benefits generated by investing in the use of physical metrics and financial valuation methods, to compare and contrast the cost and wider benefits of different solutions to managing rainwater and groundwater in line with business priorities. This will allow them to make better-informed decisions based not just on financial capital but also the natural, social, and human capital benefits generated.

The global problems (groundwater overexploitation-induced negative impacts of pumping, pollution, resource degradation, conflicts, etc.) and the overviews on international scenarios presented earlier clearly indicate that RWH for effective groundwater resources management is virtually non-existent or still in its infancy in many countries. Factors that explain these shortcomings include lack of awareness and understanding, negligence, selfish behavior, conflicts of interests, and poor governance. Progress towards achieving groundwater management goals is very difficult due to water scarcity, related groundwater resource externalities, constraints inherent to the local groundwater resources, difficult socioeconomic conditions, and an adverse political setting in some areas.

Billions of people still lack safe water and sanitation facilities, water pollution is worsening, governances are weak and fragmented, and political, institutional and administrative rules, practices and processes are inadequate in many countries. Agriculture places enormous stress on water, and institutional and human capacity is insufficient across the water sector, particularly in the least developed countries. It is very desirable to monitor water resources more with reliable data, adapted to country contexts, to create multi-stakeholder partnerships and strengthen regional integration, eliminate inequalities, ensure public participation, and develop the necessary long-term human and institutional capacity. Better groundwater governance is an important aspect in this context. For effectiveness of this approach, it is desirable that groundwater managers be more aware of and better informed about common goals and priorities, and integrate their responsibilities and activities with groundwater institutions, by encouraging community participation.

To make RWH useful for groundwater management, it may be more meaningful to adopt the positive examples of RWH, simultaneously constructively analyzing the cases where RWH for groundwater resources management is deficient, missing, or failing.

The water sector has to recognize that decisions by people determine water use. Decision-making approaches within the water sector need to integrate the technological aspects of water resources management with other socioeconomic sectors. Because many problems, such as the structural reduction of available resources in groundwater overabstraction areas and/or the linked crises of water, climate change, energy, food supply, and financial markets cannot be solved within the narrow limits of a single sector, leaders in government, the private sector, NGOs, and civil societies must learn to recognize water's role in obtaining their objectives in practical reality. It is a challenge to groundwater professionals and other sector specialists to be open-minded and recognize situations and problems being addressed on a wider platform than offered by their own sector, for the benefit of the public. The opportunities to turn this into reality may vary from country to country.

References

Datta, P.S. (2013). Groundwater vulnerability to changes in land use and society in India. Understanding Freshwater Quality Problems in a Changing World, Proceedings of H04, IAHS-IAPSO-IASPEI Assembly, July, Gothenburg, pp. 345–352.

Datta P.S. (2015). Traditional Rainwater Harvesting at Crossroads – Time to Seek Ethical Opportunities for Reform. Proceedings of the Aqua Foundation IX World Aqua International Congress, Reviving Traditional Water & Environment Conservation Techniques, November 26–27, New Delhi, pp. 93–115.

Glossary of Terms

Abstraction The process of taking water from a source, either temporarily or permanently. The volume of water taken during this process per unit of time. *Synonyms: withdrawal, extraction.*

Air gap A vertical space between a water or drain line and the flood level of a receptacle used to prevent backflow or siphonage from the receptacle in the event of negative pressure or vacuum.

Alluvium Deposits of clay, silt, sand, gravel, or other particulate material that have been deposited by a stream or other body of running water in a streambed, on a flood plain, on a delta, or at the base of a mountain.

Aquifer A geological formation that is water bearing. A geological formation or structure that stores and/or transmits water, such as to wells and springs. Use of the term is usually restricted to those water-bearing formations capable of yielding water in sufficient quantity to constitute a useable supply.

Aquifer (confined) Soil or rock below the land surface that is saturated with water. There are layers of impermeable material both above and below it and it is under pressure so that when the aquifer is penetrated by a well, the water will rise above the top of the aquifer.

Aquifer (unconfined) An aquifer whose upper water surface (water table) is at atmospheric pressure, and thus is able to rise and fall.

Arid region A region of low precipitation, characterized by a severe lack of available water, to the extent of hindering or even preventing the growth and development of plant life. In agriculture, this term is used to indicate extremely dry areas where without irrigation no crops can be grown.

Artesian water Ground water that is under pressure when tapped by a well and is able to rise above the level at which it is first encountered. It may or may not flow out at ground level. The pressure in such an aquifer commonly is called artesian pressure, and the formation containing artesian water is called an artesian aquifer or confined aquifer.

Artificial recharge Any man-made scheme or facility that adds water to an aquifer.

Backflow Flow of water in a pipe or water line in a direction opposite to normal flow.

Backflow preventer A device or system installed in a water line to stop backflow from a non-potable source.

Water Harvesting for Groundwater Management: Issues, Perspectives, Scope, and Challenges, First Edition. Partha Sarathi Datta.

Backwater The wastewater from toilets and kitchen sinks.

Base flow Sustained flow of a stream in the absence of direct runoff. It includes natural and human-induced stream flow. Natural base flow is sustained largely by groundwater discharges.

Bedrock The solid rock beneath the soil and superficial rock. A general term for solid rock that lies beneath soil, loose sediments, or other unconsolidated material.

Bore well Small-diameter wells, which are generally deeper than open wells.

Brackish water Water containing dissolved solids in a concentration between 1000 and 10 000 $mg\,L^{-1}$.

Buffer To shift pH to a specific value.

Building footprint The area of a building on the ground.

Capillary action The means by which liquid moves through the porous spaces in a solid, such as soil, plant roots, and the capillary blood vessels in our bodies due to the forces of adhesion, cohesion, and surface tension. Capillary action is essential in carrying substances and nutrients from one place to another in plants and animals.

Catchment Total area having a common outlet for its surface water discharge. *Synonyms: watershed, river basin.*

Catchment area The area from which runoff flows into a river, reservoir, etc.

Check dam Small dam constructed in a gully or other small watercourse to decrease stream flow velocity, minimize channel scour, and promote deposition of sediment.

Chlorination The use of chlorine for the treatment of water, sewage or industrial wastes for disinfection or oxidation.

Cistern An above- or below-ground tank used to store water, generally made of galvanized metal, fiberglass, ferrocement or concrete.

Climate Synthesis of local weather conditions, presented as statistics (mean values, variances, probabilities of extreme values, etc.) over a long period of time (usually taken as 30 years).

Climate change Long-term modification of the climate (manifested by a change in the long-term statistical properties of climatic variables, in particular in averages and/or variability).

Contamination To introduce a substance that would cause the concentration of that substance to exceed the maximum contaminant level and make the water unsuitable for its intended use.

Depletion Reduction of the stored volume of groundwater in an aquifer (also, reduction of the stored water volume of any other component of the local hydrological cycle).

Discharge The volume of water that passes a given location within a given period of time. Usually expressed in cubic feet per second.

Disinfection A process in which pathogenic (disease-producing) bacteria are killed by use of chlorine or physical processes.

Diverter A mechanism designed to divert the first flush of rainwater from entering the cistern.

Drainage area The drainage area of a stream at a specified location is that area, measured in a horizontal plane, which is enclosed by a drainage divide.

Drainage basin Land area where precipitation runs off into streams, rivers, lakes, and reservoirs. It is a land feature that can be identified by tracing a line along the highest elevations between two areas on a map, often a ridge. Large drainage basins, such as the area that drains into a large river, contain thousands of smaller drainage basins. Also called a "watershed."

Dug wells Traditional large-diameter wells. Defined precisely as pits excavated in the ground until the water table is reached, supported on the sides by roller compacted concrete/bricks/stones/walls; diameters could vary from 0.6 m onwards.

Ecosystem A dynamic complex of plant, animal, and micro-organism communities and their non-living environment interacting as a functional unit.

Erosion The loss of topsoil that occurs as a result of runoff.

Evaporation The process of liquid water becoming water vapor, including vaporization from water surfaces, land surfaces, and snow fields, but not from leaf surfaces.

Evapotranspiration The sum of evaporation and transpiration.

Filtration The process of separating particles of 2 μm or larger in diameter from water by means of a porous substance such as a permeable fabric or layers of inert material housed in a media filter or removable cartridge filter.

First flush Generally, the first 50 L of rainwater per 1000 square feet of roof surface that is diverted due to the potential for contamination.

Floodplain A strip of relatively flat and normally dry land alongside a stream, river, or lake that is covered by water during a flood.

Flow rate The quantity of water which passes a given point in a specified unit of time, expressed in liters per minute.

Flowing well/spring A well or spring that taps groundwater under pressure so that water rises without pumping. If the water rises above the surface, it is known as a flowing well.

Force breaker An extension of the fill pipe to a point 1″ above the bottom of the cistern, which dissipates the pressure of incoming rainwater and thus minimizes the stirring of settled solids.

Fossil groundwater Stored groundwater that has entered a rock formation during a remote period of time (under climatic and/or geological conditions different from those of the present day) and is not renewed under present-day conditions. *Synonym: non-renewable groundwater.*

Fresh water Water that contains less than 1000 $mg\,L^{-1}$ of dissolved solids; generally, more than 500 $mg\,L^{-1}$ of dissolved solids is undesirable for drinking and many industrial uses.

Gray water The waste water from residential appliances or fixtures except toilets and kitchen sinks.

Groundwater The water retained in the intergranular pores of soil or fissures of rock below the water table. (i) Water that flows or seeps downward and saturates soil or rock, supplying springs and wells. The upper surface of the saturated zone is called the water table. (ii) Water stored underground in rock crevices and in the pores of geological materials that make up the Earth's crust.

Groundwater, confined Groundwater under pressure significantly greater than atmospheric, with its upper limit being the bottom of a bed with hydraulic conductivity distinctly lower than that of the material in which the confined water occurs.

Groundwater draft Quantity of groundwater withdrawn from groundwater reservoirs.

Groundwater level Elevation to which groundwater will or does rise in a piezometer connected to a point in the groundwater domain. It is a time-dependent variable, varies from point to point within the groundwater domain, and indicates the potential energy of groundwater in any point considered (in meters of water column relative to a selected topographic reference level). *Synonyms: piezometric level, piezometric head, hydraulic head, groundwater hydraulic potential.*

Groundwater recharge Inflow of water to a groundwater reservoir from the surface. Infiltration of precipitation and its movement to the water table is one form of natural recharge. Also, the volume of water added by this process.

Groundwater table Surface defined by the phreatic levels in an aquifer (i.e. surface of atmospheric pressure within an unconfined aquifer).

Groundwater, unconfined Water in an aquifer that has a water table that is exposed to the atmosphere.

Hardness A characteristic of groundwater due to the presence of dissolved calcium and magnesium, which is responsible for most scale formation in pipes and water heaters.

Humid region Region characterized by a surplus of precipitation over potential evapotranspiration. In other words, the available water in the area is sufficient to sustain the growth and development of plant life without irrigation.

Hydrological cycle The continual exchange of water from the atmosphere to the land and oceans and back again. The cyclic transfer of water vapor from the Earth's surface via evapotranspiration into the atmosphere, from the atmosphere via precipitation back to Earth, and through runoff into streams, rivers, and lakes, and ultimately into the oceans.

Impermeable layer A layer of solid material, such as rock or clay, which does not allow water to pass through.

Infiltration Flow of water from the land surface into the subsurface.

Infiltration gallery Man-made construction to tap groundwater and conduct it to the surface without the need for external energy (the system is based on flow by gravity).

Leaching Process by which soluble materials in the soil, such as salts, nutrients, pesticide chemicals or contaminants, are washed into a lower layer of soil or are dissolved and carried away by water.

Leaf screen A mesh installed over gutters and entry points to downspouts to prevent leaves and other debris from clogging the flow of rainwater.

Masonry A wall or other structure made using building blocks such as bricks or stone with binding materials such as cement or lime.

Micron A linear measure equal to one millionth of a meter, or 0.00003937 in.

Milligrams per liter (mg L^{-1}) A unit of the concentration of a constituent in water or waste water. It represents 0.001 g of a constituent in 1 L of water. It is approximately equal to one part per million (ppm).

Million gallons per day (Mg d^{-1}) A rate of flow of water equal to 133 680.56 cubic feet per day, or 1.5472 cubic feet per second, or 3.0689 acre-feet per day. A flow of one million gallons per day for one year equals 1120 acre-feet (365 million gallons).

Non-point source pollution Pollution discharged over a wide land area, not from one specific location. These are forms of diffuse pollution caused by sediment, nutrients, organic, and toxic substances originating from land use activities, which are carried to lakes and streams by surface runoff. Non-point source pollution is contamination that occurs when rainwater, snowmelt, or irrigation washes off plowed fields, city streets, or suburban backyards. As this runoff moves across the land surface, it picks up soil particles and pollutants, such as nutrients and pesticides.

Non potable water Water intended for non-human consumption purposes, such as irrigation, toilet flushing, and dishwashing.

Open well Same as dug well. These wells were kept open in earlier times for manual withdrawal of water. Today, with electrical or diesel/patrol pumps, they can be fully covered.

Organic matter Plant and animal residues, or substances made by living organisms. All are based upon carbon compounds.

Parts per billion The number of "parts" by weight of a substance per billion parts of water. Used to measure extremely small concentrations.

Parts per million The number of "parts" by weight of a substance per million parts of water. This unit is commonly used to represent pollutant concentrations.

Pathogen An organism which may cause disease.

Per capita use The average amount of water used per person during a standard time period, generally per day.

Percolation (i) The movement of water through the openings in rock or soil. (ii) The entrance of a portion of the stream flow into the channel to contribute to groundwater replenishment.

Permeability The ability of a material to allow the passage of a liquid, such as water through rocks. Permeable materials, such as gravel and sand, allow water to move quickly through them, whereas unpermeable material, such as clay, does not allow water to flow freely.

pH A logarithmic scale of values of 0–14 that measure hydrogen ion concentration in water which determines whether the water is neutral (pH 7), acidic (pH 0–7) or basic (pH 7–14).

Point-source pollution Water pollution coming from a single point, such as a sewage outflow pipe.

Porosity A measure of the water-bearing capacity of subsurface rock. With respect to water movement, it is not just the total magnitude of porosity that is important, but the size of the voids and the extent to which they are interconnected, as the pores in a formation may be open, or interconnected, or closed and isolated. For example, clay may have a very high porosity with respect to potential water content, but it constitutes a poor medium as an aquifer because the pores are usually so small.

Potable water Water of a quality suitable and safe for human drinking.

Precipitation Rain, snow, hail, sleet, dew, and frost.

Pressure tank A component of a plumbing system that provides the constant level of water pressure necessary for the proper operation of plumbing fixtures and appliances.

Rainwater harvesting The principle of collecting and using precipitation from a catchment surface.

Recharge The process of surface water (from rain or reservoirs) being added to the groundwater aquifer. For instance, rainfall that seeps into the ground.

Replenishable groundwater The portion of precipitation which, after infiltration, percolates down and joins the groundwater reservoir.

Reservoir A pond or lake built for the storage of water, usually by the construction of a dam across a river.

Resilience Ability of a system to recover from an unsatisfactory state (in particular, the ability to remain in or return to a state of dynamic equilibrium).

River A natural stream of water of considerable volume, larger than a brook or creek.

Roof washer A device used to divert the first flush of rainwater from entering a cistern.

Runoff The term applied to the water that flows away from a surface after falling on the surface in the form of rain. (i) Part of the precipitation, snow melt, or irrigation water that appears in uncontrolled surface streams, rivers, drains or sewers. Runoff may be classified according to speed of appearance after rainfall or melting snow as direct runoff or base runoff, and according to source as surface runoff, storm interflow, or groundwater runoff. (ii) The total discharge described in (i), above, during a specified period of time. (iii) Also defined as the depth to which a drainage area would be covered if all the runoff for a given period of time were uniformly distributed over.

Saline water Water that contains significant amounts of dissolved solids. Fresh water (<1000 ppm); slightly saline water (1000–3000 ppm); moderately saline water (3000–10 000 ppm); highly saline water (10000–35 000 ppm).

Salt water intrusion, sea water intrusion Invasion of salt water (usually sea water) into a body of fresh water (either a surface water or a groundwater body).

Sediment Usually applied to material in suspension in water or recently deposited from suspension. In the plural, the word is applied to all kinds of deposits from the waters of streams, lakes, or seas.

Sedimentary rock Rock formed of sediment, specifically (i) sandstone and shale, formed of fragments of other rock transported from their sources and deposited in water; and (ii) rocks formed by or from secretions of organisms, such as most limestones. Many sedimentary rocks show distinct layering, which is the result of different types of sediment being deposited in succession.

Sedimentation The process in which solid suspended particles settle out (sink to the bottom) of water, frequently after the particles have coagulated.

Seepage (i) The slow movement of water through small cracks, pores, interstices, etc. of a material into or out of a body of surface or subsurface water. (ii) The loss of water by infiltration into the soil from canals, ditches, laterals, watercourses, reservoirs, storage facilities, or other body of water, or from a field.

Stream A general term for a body of flowing water; natural wate course containing water at least part of the year. In hydrology, it is generally applied to the water flowing in a natural channel as distinct from a canal.

Stream flow The water discharge that occurs in a natural channel. A more general term than runoff, stream flow may be applied to discharge whether or not it is affected by diversion or regulation.

Surface water Water that is on the Earth's surface, such as in a stream, river, lake, or reservoir.

Total dissolved solids A measure of the mineral content of water supplies.

Transpiration Process by which water that is absorbed by plants, usually through the roots, is evaporated into the atmosphere from the plant surface, such as leaf pores. See "Evapotranspiration."

Unsaturated zone The zone immediately below the land surface where the pores contain both water and air, but are not totally saturated with water. These zones differ from an aquifer, where the pores are saturated with water.

Virtual water Water used in the production of goods or services. *Synonym: embedded water.*

Water cycle The circuit of water movement from the oceans to the atmosphere and the Earth and return to the atmosphere through various processes such as precipitation, interception, runoff, infiltration, percolation, storage, evaporation, and transportation.

Water footprint The total volume of fresh water used to produce the goods and services consumed by an individual or community or produced by a business.

Water pollution The addition of harmful or objectionable material causing an alteration of water quality.

Water quality A term used to describe the chemical, physical, and biological characteristics of water, usually in respect to its suitability for a particular purpose.

Water table The top of the water surface in the saturated part of an aquifer.

Water use Water that is used for a specific purpose, such as for domestic use, irrigation, or industrial processing. Water use pertains to human interaction with and influence on the hydrological cycle, and includes elements such as water withdrawal from surface and groundwater sources, water delivery to homes and businesses, consumptive use of water, water released from wastewater treatment plants, water returned to the environment, and in-stream uses, such as using water to produce hydroelectric power.

Watershed The land area that drains water to a particular stream, river, or lake. It is a land feature that can be identified by tracing a line along the highest elevations between two areas on a map, often a ridge. Large watersheds, like the Mississippi River basin, contain thousands of smaller watersheds.

Well A man-made construction, usually of vertical cylindrical form, used to obtain access to groundwater and abstract it by lifting it to the surface.

Wetlands Areas of marsh, fen, peatlands or water, natural or artificial, permanent or temporary, with water that is static or flowing, fresh, brackish or salt, including areas of marine water less than 6 m deep at low tide.

Withdrawal Water removed from a surface or groundwater source. *Synonym: abstraction.*

Index

Water Harvesting for Groundwater Management: Issues, Perspectives, Scope, and Challenges, First Edition. Partha Sarathi Datta.